Tales of Science

By Joan S. Wagner

© January 1, 2021

Topical Review Book Company
P.O. Box 328
Onsted, MI 49265

topicalrbc@aol.com
Focusonlearning1@aol.com

www.topicalrbc.com

Table of Contents

A little Astronomy..1-21
Emilia receives a telescope from her grandmother for her birthday. She has many questions about the universe and even builds a Moon Clock with the help of her grandmother, who is a science teacher.
In this story, moon phases, the make-up of the universe and how the motion of Earth is related to the seasons and time is explored.

Earth Shapes Up...22-37
Students attending the Martin Luther King Middle School were being dropped off at their school at 6:00 AM for a field trip to the Adirondacks. They were divided into groups of four led by a parent or teacher.
As they hike the Adirondacks, they learn about Earth processes and geology.

The Big Competition..38-53
The students in Mr. Novak's science class were very excited about the climate change competition. Mr. Novak was the type of teacher who always searched for new ways to motivate his students about learning. When this competition was announced online, he thought it would be both fun and educational because it was based on issues facing our planet now. Students were challenged to produce an educational video about climate change. The winning top 20 teams get a free trip to Washington, DC for a weekend of science activities.
The focus of this story is to learn about the science of climate change.

It's Your Planet: There is No Planet B: An Earth Day Celebration......................54-76
The students in Ms. Gomez's class were busy preparing their exhibits for Earth Day.
"I can't believe that Earth Day is almost here. It seems just like yesterday when Ms. Gomez told us about this project," said Griffin.
"That was in September, "replied Nikko, "And now it is April. The year certainly has flown by."
Students develop projects that identify problems facing our planet and provide solutions.

Farrah's Electric Birthday Party...77-89
Farrah was very excited because it was her 12th birthday party. She and her brother Eli were helping their mom decorate for the party. Guests would start arriving in a few hours. "Let's blow up these balloons and stick them to the walls," said their mom, Lisa. Dad, JD was helping too.
In this themed birthday party, the guests learn about electricity and magnetism.

The Big Debate: Analog (Vinyl) vs. Digital Music..90-107
Hunter and Dylan were getting together to listen to some music with their friends, Chessie and Kiley. Dylan was excited about the new vinyl records he purchased.
"Vinyl sound is so much better than the sound from CDs because it uses an analog storage system instead of digital," said Dylan to Hunter.

"The debate between analog vs. digital music has been around for quite a while," said Hunter.
"Quite honestly, I don't really hear a difference."
This story uses digital vs analog to learn about wave energy and the properties of waves and how information today is electronically stored.

It's Elementary..108-130
Allison was very excited about her chemistry set that she got for her birthday.
"Let's explode something," said her friend Jordyn as they unpacked the chemistry set.
"Everyone always wants to explode things, I want to test out the chemicals and learn about their properties. If I am going to be a chemist, I need to start early, said Allison.
"I bet you change your mind when you get to college,' said Jordyn.
"I doubt it," replied Allison.
Using a chemistry set, the girls explore the properties of matter.

The Case of the Ball that Would Not Bounce..131-149
Two girls find a ball that will not bounce. They decide to figure out why it will not bounce.
This story deals with forces and energy.

The Case of the Dent in Mom's Car..150-163
Dad notices a dent on the top of Mom's car. Mom has no idea how it happened. Her children, Walter and Evie decide to investigate the cause of the dent.
As they investigate they learn about forces and motion.

Staying Alive..164-181
Lauren, Gavin, Grant, Hillary and Brady were walking to the park together. Hillary was baby-sitting for her little sister, Olive who was dragging her stuffed beagle along.
"Stop dragging your beagle. You will hurt it," said Hillary to her little sister, who continued to drag it.
"Come on Hillary, how can you hurt a stuffed animal. It's not alive," said Grant.
This story is about what it means to be alive.

We are Alike and Different..182-208
Fraternal twins Eliza and Oliver become curious over how they are alike and different. Their curiosity was perked after reading a book given to them by their mother.
Using literature and games, the twins learn about heredity.

Life is A'Changin'...209-241
Drew, Noah, Crystal and Caitlin were excited about their field trip to the Science Center because of the dinosaur exhibit.
"I can't wait to see the dinosaurs," said Noah.
"My friend Noe was there and she said they are very life-like. There is even a mama dinosaur with her babies hatching," said Crystal.

During a field trip to a Museum featuring dinosaurs, students learn how living things change and adapt to their environment.

Project Pond/Wetland..242-261
It was the first day of school and Mr. Hull was very excited about utilizing the pond, wetland and butterfly field students, faculty and the community built for the Middle School. He and other science teachers in the building had collaborated on the development of an extensive curriculum for the Middle School during the summer. The students at Mayville Middle School were as excited as the teachers.
This story is about ecosystems and how energy is cycled through it.

Thanks..262

A Little Astronomy

Don't be Fazed by Phases of the Moon

Emilia was very excited as she opened the birthday present her grandmother, Jenna, had just given her.

"A telescope! This is what I always wanted!"

Jenna smiled at Emilia's excitement. As a science teacher, she had been answering Emilia's many questions for a while now. Over time, she had noticed that Emilia really liked to ask questions about stars and planets. After last year's present, a book about the constellations that glowed in the dark, Emilia had given her a very enthusiastic phone call to thank her.

"Can we look at the stars today?" Emilia asked.

"Of course! Around this time of the year, we don't have to wait long for the sun to set either. Tonight, it will set around 5:30 this evening."

Emilia's excited face turned to one of confusion. "Okay, but why can I see the moon now, even though it isn't dark?"

"Many people think that you can see the moon only after the sun has set, but guess what? The moon rises at different times of the day, just like the sun, depending on its phase."

"Oh! Okay, I understand now. Did you know that we learned about the moon phases in class?"

"Yes. You told me the same day you learned about them in class," Jenna said with a smile. "What are they again?"

Emilia faced her grandmother and took a deep breath. "The main phases are: full, new moon, waxing crescent, first quarter, waxing Gibbous, full moon, waning gibbous, last quarter and waning crescent."

"Great job!"

Emilia beamed, proud of herself!

"Did you know that each phase of the moon rises, sets and is highest above the horizon at specific times? In fact, you can use the phases to tell time. For example, at the equinox, when the time between sunrise and sunset is exactly twelve hours, the moon rises at 9 A.M. and sets at 9 P.M. It is highest above the horizon at 3:00 P.M." Jenna said.

Thinking for a moment, Emilia responded. "So, does that mean that you can see the Moon almost all day?"

Jenna thought about an exercise she used in one of her classes to teach. "How about we make a Lunar Clock? That should help answer your question."

"What is a Lunar Clock?" Asked Emilia looking very puzzled.

"You will see," answered her Grandma.

Jenna got up and asked her son for some materials to make the moon clock. She needed, card quality paper, scissors and a metal fastener. After instructing her granddaughter, Emilia assembled the clock.

Her grandmother provided her with a chart to record data, similar to what she used with her students.

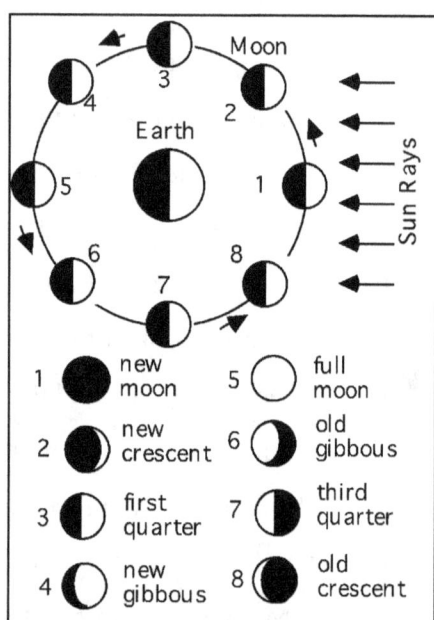

Suddenly, Emilia's little brother, Zach, came downstairs and wanted to get involved with whatever was going on between his big sister and grandmother. "What are you doing?"

Jenna turned to her youngest grandchild. "We are making a moon clock. It allows you to tell time by the phase of the moon."

"How does it work? How does it work?" Zach wondered, bouncing slightly.

Jenna patted the seat beside her and Zach sat down, across from his sister at their table. "Well, it's easy to see that the new moon rises and sets with the sun," Jenna told them. Their grandmother nodded and demonstrated this by placing the new moon on the east horizon. "The moon rises in the east, just like the sun. Notice the time. It's 6:00 in the morning." Jenna then placed the new moon on the west horizon. The time read 6:00 in the evening. "Since the sun must point to the back of the moon, the new moon will always rise and set with the sun. Okay, now that we know this, I want you to make predictions for other times."

Emilia's birthday party quickly became a science lesson as she and her brother filled in times in the lunar phase clock chart. After they finished, they showed the chart to Jenna.

"Do you know what phase the moon is in tonight?" Asked Jenna.

"Well, I can only see half of the moon and yesterday, I saw a little less than that. So, that must mean that it is the first quarter because it's been getting bigger each night," Emilia answered after thinking her answer over.

"Correct," Jenna responded impressed with her granddaughter's knowledge. "But, the Moon is not really getting any bigger. It just looks that way because it is reflecting more light from the sun due to the angle by which we are viewing it. In actuality, half of the moon is always illuminated by the sun, but you would need to travel into space to see that."

What follows are the directions for making a Moon Clock and a template to construct a Moon Clock. The reader can try what Emilia did. Check the end of the story to see how you did in completing the chart.

Materials: card quality paper, metal fastener, scissors

1. Copy Moon Clock Face and Dial onto card quality paper.
2. Cut out Moon clock Face with sun attached and cut out the dial.
3. Attach the dial to the face by inserting the fastener through the black dot on top of the arrow pointing to the meridian and the black circle in the middle of the clock.
4. Place the new Moon on the east horizon. All phases of the Moon rise on the east horizon.
5. The triangle points to the time it rises.
6. When the Moon phase is above the meridian, it is highest in the sky. The arrow points to the meridian. It is the sky above your head.
7. Turn the clock clockwise to determine the time each Moon phase, rises, is highest in the sky and sets. Record your answers on the Moon Chart.

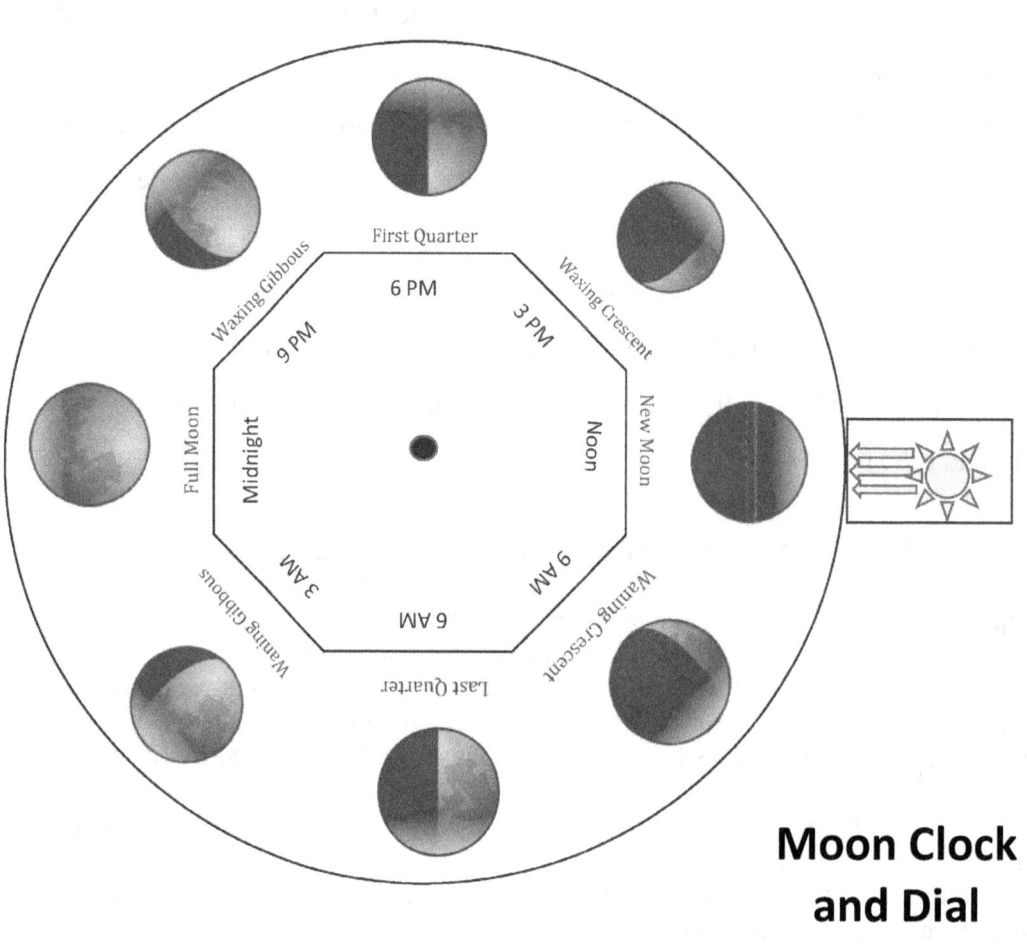

Moon Clock and Dial

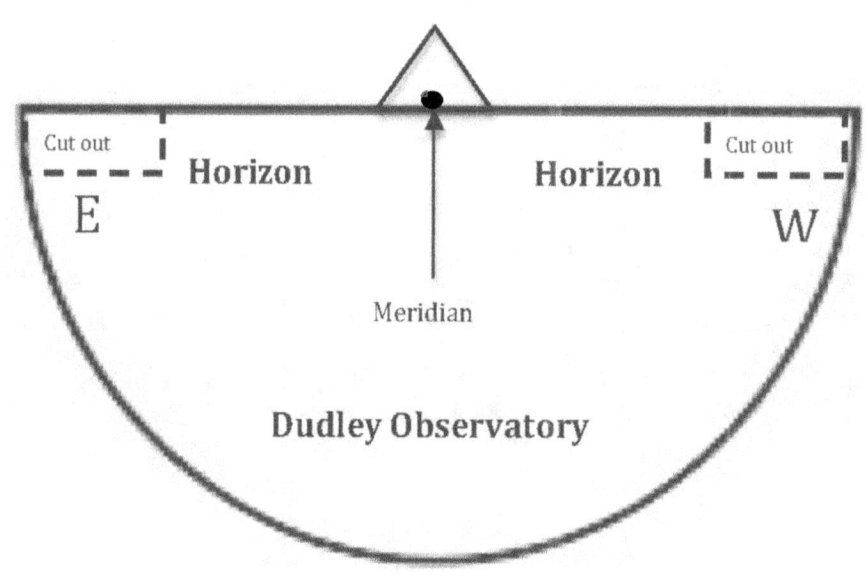

Dudley Observatory

A Little Astronomy, <u>**Tales of Science**</u> by Joan S. Wagner

MOON PHASES AND TIME OF DAY CHART

After completing the Moon clock, use the dial to complete the chart. Start with the New Moon and turn the wheel to record the information below. Provide the time the Moon rises, sets and is highest in the sky (at meridian) and place a check to tell if it is in the east or west sky.

PHASE	RISES	IN EASTERN SKY	HIGHEST IN SKY	SETS	IN WESTERN SKY
NEW					
WAXING CRESCENT					
FIRST QUARTER					
WAXING GIBBOUS					
FULL					
WANING GIBBOUS					
LAST QUARTER					
WANING CRESCENT					

A Little Astronomy, <u>Tales of Science</u> by Joan S. Wagner

It's in the Stars

"It's getting dark, Grandma! Can we please, please, try out my new telescope?" Emilia asked later that evening.

"Can I look too?" Zach asked, worried that he wouldn't be able to see.

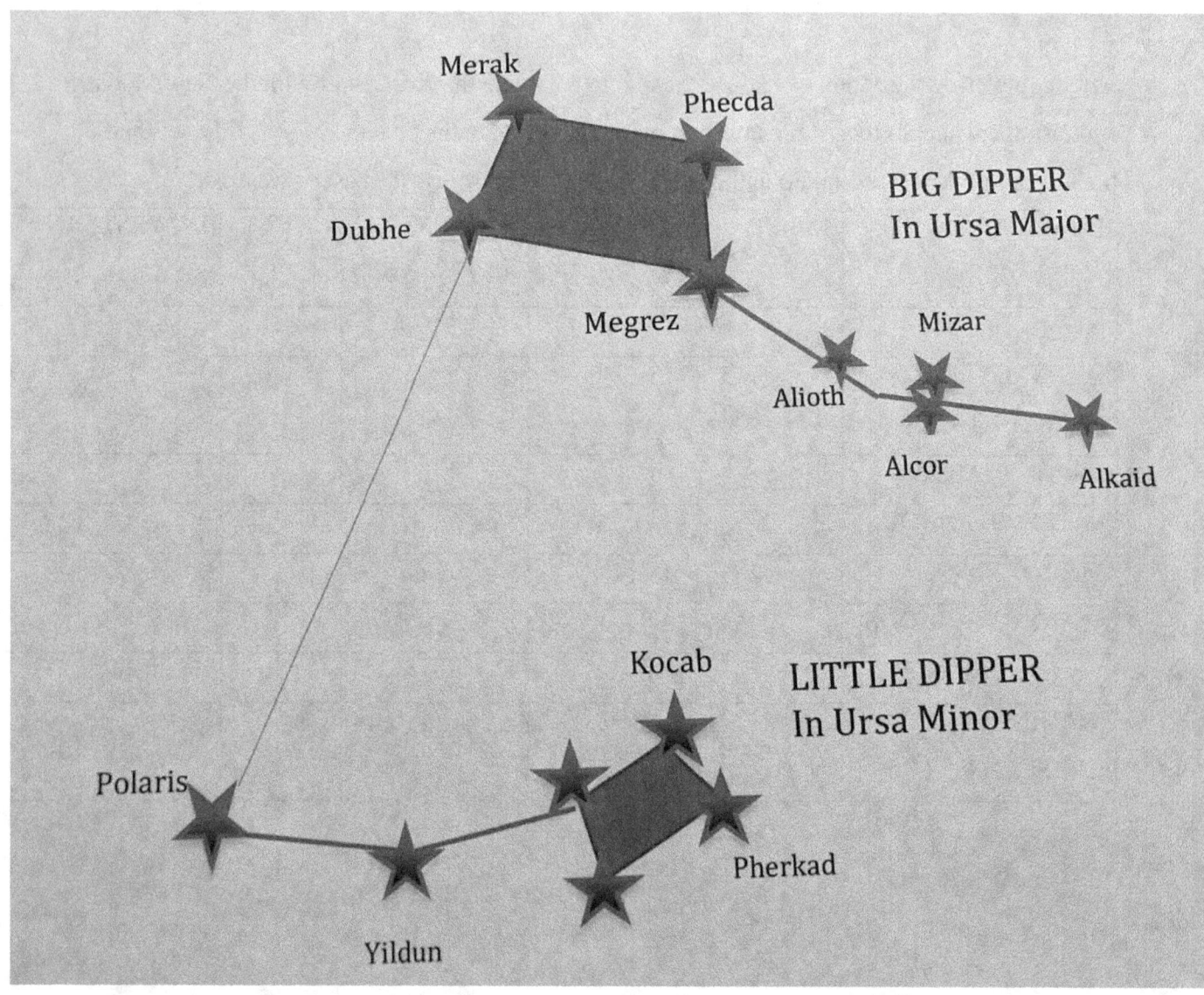

It became a family outing, as Emilia's parents joined them outside to view the evening sky. Patches of snow covered the family's deck from a recent fall. Jenna pointed out Venus, Mars and Jupiter.

"How can you find things so quickly in the night sky, grandma?" Emilia asked, looking at her in awe.

"Well, it's sort of like trying to find a location on a roadmap," she responded. "Can you see the Big Dipper (Ursa Major)?"

Emilia looked for a moment and pointed to it. "There. It's facing the north."

"Good job! So, if I told you that the Big Dipper is above the Little Dipper, and that its ladle points in the opposite direction of the Big Dipper's ladle, can you find it?"

Zach, who was standing close beside his dad, was carefully listening. "I see it! I see it, grandma!"

Emilia quickly found it after her younger brother.

"Good job, kids! Now. What about Polaris, the North Star, if I told you that the end of the handle of the Little Dipper points towards it?"

After they found it, Zach asked why the groups of stars were called 'Little Dipper' and 'Big Dipper.'

"Well, we know that there aren't real spoons in the evening sky, but a long time ago, people looked up at the sky and created shapes and stories to help them identify the stars in the night sky. These shapes are called constellations. The Big Dipper is part of the constellation, Ursa Major or 'Big Bear' while the Little Dipper is part of the constellation Ursa Minor or 'Little Bear,'" their dad explained.

"Yes, it is an example of what people did in the past to help them identify the stars. If you really use your imagination, you can see a bear in the night sky," Jenna added. "The early Greeks even made up stories about the star patterns they saw in the evening sky. According to their legend, the Big Bear, Ursa Major, was a beautiful woman named Callisto. She was turned into a bear by Zeus, the god of the sky, to protect her from his jealous wife, Hera. One day, Callisto's son, Arcas, was hunting in the woods where he came across a bear. Arcas raised his spear towards the bear, not knowing that it was his mother. Zeus, who was watching from above, acted quickly to save his beloved Callisto. He turned Arcas into a bear and hoisted them into the sky by their tails. This toss stretched out their tails, giving them the appearance, they now have in the night sky."

"Wow! And each constellation has a story? I want to read about all of them!" Emilia thought out loud excitedly.

"Does our sun belong to a constellation?" Zach asked.

"No," Jenna responded. "Remember, Zach, our planet revolves around the sun. The positions of the stars change relative to us. There are twelve constellations that our sun passes through which we call the Zodiac."

"So, you could say our sun belongs to twelve constellations?" he asked.

"Sort of. Think about this. Suppose you were able to move to a planet that revolves around a different star. Scientists call planets that orbit around other stars exoplanets. Scientists have used the space telescope, Kepler, to study exoplanets, though it has been retired from use in 2018. So far, thousands have been found!" Jenna explained.

"And that means that when you look up at the night sky, what you see depends on where you are, right?" Emilia wondered.

"Right. The time of the year is also a factor."

"And all the stars we see, including our sun are part of the Milky Way Galaxy," said Emilia.

"Right again," said Emilia's grandmother, "but our galaxy is just one of billions of other galaxies."

"Or billions and billions," added Zach.

"Hard to even think of numbers that big," said Emilia.

Emilia and her family were enjoying the evening identifying stars when they were suddenly treated to a bonus. The International Space Station (ISS) went by. It looked like a moving dot of light in the night sky and took less than a minute to clear the horizon. It was launched in 1998.

"Wow," that was neat," said Zach.

Microgravity

"What is the International Space Station," asked Emilia.

"You can think of it as an orbiting laboratory in which experiments in all of the science disciplines are carried out. What is neat about this lab is that there is microgravity. Scientists from many countries are learning a lot about how living things and different types of matter respond to microgravity," said Emilia grandmother.

"What do you mean by microgravity?" Asked Emilia.

Emilia's grandmother went on to explain microgravity. She said, "Microgravity is when things seem to be weightless. Anything that falls toward Earth has microgravity during the period of free fall. If you throw a baseball, gravity will cause it to curve down and strike Earth. The International Space Station is actually in free fall toward Earth. However, its speed of 17,500 mi/hr, allows for the curve of its fall to match the curve of our planet so as it falls, it falls around the curve of our planet. Therefore, it can never hit our planet. This is the same reason out planet doesn't fall into the sun and the moon does not fall into Earth."

"Viewing the night sky is really interesting," said Emilia. She thanked her grandmother for all of the interesting information about microgravity.

The family stayed out a while longer and looked for stars Jenna suggested through Emilia's new telescope.

Meteoroids, Asteroids, Comets and Planets

The next day, Emilia had tons more questions for her grandmother. She had stayed up later than she was supposed to, reading a book on outer space she got from her parents the day before. It was great that her grandmother had stayed; she could answer all the new questions she had.

Today, Emilia had questions on meteorites. She had read about the meteorite that struck Russia. This occurrence and bedtime reading, raised many questions. At breakfast, Emilia sat beside her grandma and asked, "Where do meteorites come from? What are they?"

"Good morning to you too," Jenna said warmly. "Most meteorites come from asteroids, which you can think of as large pieces of rock that orbit our sun in the area between Mars and Jupiter. Scientists refer to this area as the asteroid belt. They are left over pieces of rock that never became part of a planet. However, I should point out that before they hit our planet, they are called meteors."

"Ok, so how do they reach Earth?" Emilia queried.

Jenna took a sip of her coffee before answering. "Sometimes there are collisions and pieces of the asteroid break off and are tossed into space. If they come close to Earth, our planet's gravity pulls it toward us. As it moves through our atmosphere, friction burns a lot of it up. That is what we call 'shooting stars.' A really large one fell in Russia a while ago, but fortunately, no one was hurt too badly."

"Grandma, you said that most meteorites come from asteroids. Where else do they come from?"

"Scientists have identified meteorites from Mars and the moon."

"Really? How is that possible? How do we know they came from Mars and the moon?"

"Excellent question. Scientists can often determine things indirectly. Believe it or not, a meteorite contains a lot of information. The gases trapped in some meteorites perfectly matched the gases known to be in Mars' atmosphere. It is this evidence that would lead a scientist to conclude that the meteorite came from Mars. Those from the moon match the geology of the rock samples astronauts brought back from the moon. On our moon, there is no atmosphere, so no gases can be trapped," she explained, before taking a long drink of coffee.

Emilia grinned. "That's cool, to use gases to determine where the meteorites came from, but why do scientists study this stuff?"

"I am glad you asked. Scientists are interested in them for a number of reasons. The main reason is they provide information about the early solar system because they have not changed much since the solar system formed. Rocks on Earth go through rock cycles. They are weathered, melted and placed under lots of pressure, which affects their form and composition. Any meteorite will have changed little in the 4.6 billion years in our solar system," Jenna told her granddaughter.

Emilia's mother placed a plate of 'hot off the griddle' pancakes in front of her and Jenna, and both thanked her before eating.

"I know something interesting about an asteroid that hit our planet," Emilia said after eating a few bites.

"Oh?" Jenna asked, biting into a piece of bacon.

She listened as Emilia told her that the dinosaurs went extinct after an asteroid struck the planet 66 million years ago. She continued, explaining that scientists found evidence to support this under the Yucatán Peninsula in Mexico. The Chicxulub crater is the result of the asteroid's impact.

Jenna smiled. "You are becoming quite the young scientist."

A minute later, Zach came to the table, a plate of pancakes held in a wobbly hand.

"Good morning, Zach," Jenna greeted.

"Good morning," he replied softly, still sleepy.

"We're talking about asteroids!" Emilia exclaimed.

"What are those?"

Jenna told him as he began to eat his breakfast.

"I learned about the asteroid belt in class last week. It separates the inner and the outer planets," he told his grandmother.

"Yes. I'm glad to hear that you were paying attention in class. Can you tell me what planets are the inner planets and which ones are the outer planets?

"I remember these! The planet closest to the sun is Mercury. Then Venus is next, followed by Earth and then Mars. Those are the inner planets. And they are called the rocky planets."

"Correct! Which one is the hottest?"

"Mercury?" He asked, unsure.

Emilia interjected, answering, "No, Venus!"

"Yes, that is right, Emilia. Do you know why?" Jenna winked at Zach who looked disappointed for not providing the right answer.

"Most people would think that the planet closest to the sun would be the warmest, but there are other things that can affect its temperature. Venus is the warmest planet because its atmosphere contains mostly carbon dioxide."

"And what does carbon dioxide have to do with the temperature of Venus?" Jenna asked her.

"Carbon dioxide is considered a greenhouse gas because it traps heat. That's why it's so hot," Emilia concluded.

"How much reading have you done," Jenna asked Emilia, surprised that she could give that much detail in her answer.

Emilia just grinned.

"You are right, though. Climate scientists think of Venus as a model for a planet with a "run-a-way greenhouse effect." Jenna finished drinking her coffee and turned to Zach. "Okay, Mr. Zach. Can you name the outer planets for me please?"

"The outer planets are Jupiter, Saturn, Uranus, Neptune, and not Pluto," he answered.

"What do you mean 'not Pluto'? Jenna asked, surprised.

Zach knew that his grandmother was just kidding. "Well, the scientists at NASA said that Pluto wasn't a planet anymore. It's a dwarf planet now because it didn't meet the definition of a planet."

"How dare they," Jenna joked, causing the two siblings to laugh. "Okay, so what is the definition of a planet?"

"I don't know that yet, Grandma!" Zach said, smiling.

"I'm going to have to go visit your science teacher," Jenna told him as she shook her head.

"A planet must meet three criteria: It must orbit the sun, have enough mass to be round and have cleared its neighborhood of other orbiting objects except its own lunar satellite(s). Not all scientists agree with this definition, but it holds for now. New discoveries might cause the definition to be modified again."

"Presently, 5 dwarf planets have been named. They are all found in the Kuiper Belt, a place located in the outer most part of our solar system. In addition to dwarf planets, the Kuiper belt contains many icy objects and is believed to be the home of comets that orbit our solar system," Jenna continued.

"I like comets," Emilia said. "My teacher made a model of a comet in lab using dry ice and rocks and soil. She even got it to form a tail!"

"I bet that was fun! I may have to do that with my students one day," Jenna replied. "Some people think of comets as dirty snowballs. As they move away from the sun, the ice vaporizes, forming a long cloud of water droplets that reflect light giving comets its tail."

Jenna wondered if her grandkids understood how the motions of the planets affected them.

"I can see that both of you have become quite knowledgeable about our solar system, but how much do you know about the planet you live on? If Earth stopped moving, would there still be gravity?"

Emilia looked unsure. "Is this a trick question, grandma?"

"No, it's not."

"OK, then there would still be gravity," Emilia said slowly. "My teacher said anything that has mass is affected by gravity. So, Earth will fall into the sun, just like the International Space Shuttle would fall onto Earth if it was not going fast enough," she answered.

Jenna smiled. She then decided to see if her grandchildren knew what an eclipse is but before she could ask it Zach beat her to it.

"What is an eclipse?"

Having already studied it in class, Emilia said, "It depends on the type of eclipse, solar or lunar."

"I wasn't asking you," said Zach.

"But, I know the answer," said Emilia.

"A full solar eclipse occurs when the Moon has moved between Earth and the sun so that its shadow is able to block the sun, while a full lunar eclipse happens when Earth is between the sun and moon and Earth's shadow blocks the sun so the moon cannot reflect any light," explained Emilia. She did a quick Google to show Zach a picture of both.

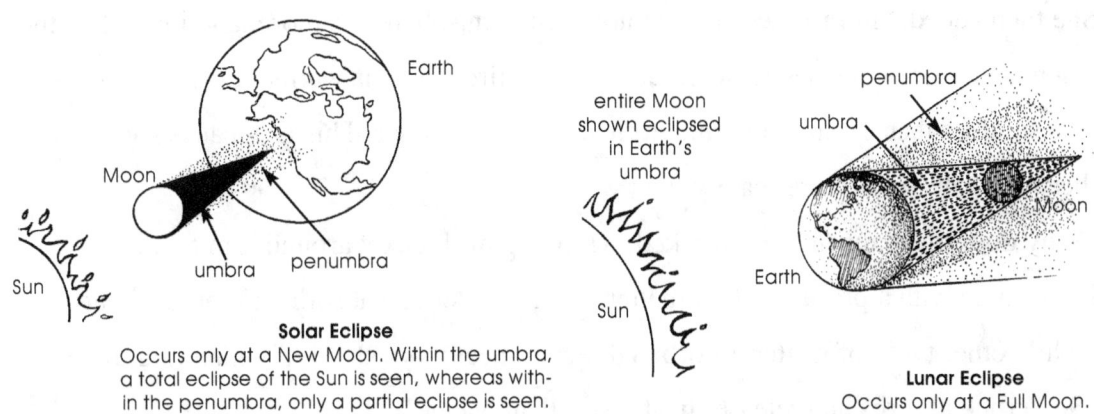

Solar Eclipse
Occurs only at a New Moon. Within the umbra, a total eclipse of the Sun is seen, whereas within the penumbra, only a partial eclipse is seen.

Lunar Eclipse
Occurs only at a Full Moon.

"And sometimes, there are partial eclipses when Earth or the Moon only partially blocks the sun," added their grandmother.

It's Seasonal!

Later that afternoon, Jenna and her grandkids were watching a TV show on climate change.

"Grandma, they didn't talk about the seasons," Zach noted.

"No, they didn't. What can you tell me about the seasons, Zach?"

"There are four: summer, fall, winter and spring," he answered. "It's hot in the summer and cold in the winter and a mix during the spring and the summer."

Jenna nodded.

Emilia was thinking. She knew from her science class that Earth is tilted 23.5 degrees from straight up and down as it orbits the sun. Because of this, when the tilt faces the sun, it is summer and it is winter when the tilt is away from the sun. Remember, Earth does not change its tilt, it is due to its position as it revolves around the sun. She explained this to Jenna and Zach.

"How does that affect the seasons?" Zach asked.

"Here's an example, Zach. The Summer Solstice is when our planet is tilted most toward the sun. We have the longest day and the most amount of sunlight striking that part of the planet. The opposite is true for the Winter Solstice."

"But aren't we farther away from the sun in the summer than in the winter? Doesn't that affect the temperature?" Emilia asked.

"No. It is the tilt toward or away from the sun that is most important. The distance is not significant to make an impact on temperature," answered Jenna.

She then added, "In fact, because the northern hemisphere has more land mass than the southern hemisphere, the average temperature of the entire planet is warmer during a northern hemisphere summer than during the southern hemisphere summer. This is because solid Earth absorbs heat and releases it more readily than water."

"Why does that happen?" Zach asked, wrapping his blanket around him tighter.

It has to do with a property of all matter. Some matter can absorb and release heat quickly, while other types of matter absorb and release heat more slowly," Jenna told them.

"So, a rock absorbs and releases heat faster than water."

"It looks like you got it, Emilia," Jenna said smiling.

"There are so many variables you have to consider when explaining scientific things. I guess when trying to figure out anything, you should look at all factors that can affect the outcome, right?" Emilia asked.

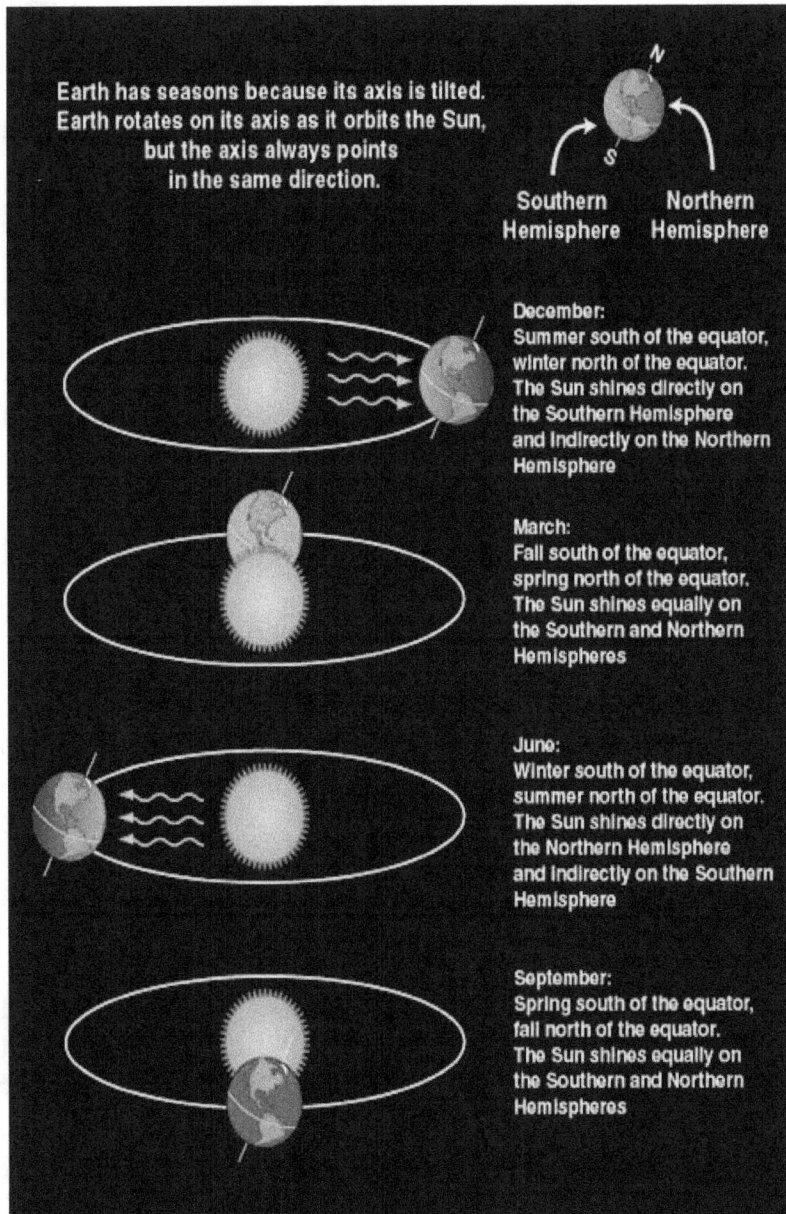

https://spaceplace.nasa.gov/seasons

"I see a budding scientist in you," Jenna replied, smiling.

"After all of our discussions, I think I have more questions than ever," exclaimed Emilia.

"Yeah," said Zach, "The more we know, the more questions we have!"

"Speaking of questions, are the days and nights ever equal in length," asked Zach.

Emilia knew the answer and quickly replied to her brother, "In the fall, about September 21, we have the autumnal equinox and in spring, about March 21, we have the vernal equinox. That is when the amount of daylight and nighttime is about equal in length."

"Oh, yeah, now I remember," said Zach.

"Now you know how scientists feel. Every time they make a new discovery, they have more questions to ask. Imagine what they thought when they began to realize our solar system formed from a disk of spinning gases that were pulled together...."

"By gravity," interrupted Emilia.

Their grandmother was going back to Saratoga Springs in a few days.

"Grandma, Zach and I really enjoy our talks about the natural world, as you call it, but mom and dad say it is time to pack our skis and head for the slopes. And I know, grandma, we can thank gravity for the ride down the trails!"

"Absolutely," said Jenna as she also picked up her skis and went out to the car.

Discussion Questions

Don't Be Fazed by Phases of the Moon

Waxing Moon brightens on the right
Clockwise it revolves day and night
Crescent, first quarter and gibbous too
Until a full Moon comes into view.
Waning Moon darkens on the right
Clockwise it revolves day and night
Gibbous, last quarter, and crescent too
Until a new Moon comes into view

1. How can you tell time by the phase of the Moon?
2. From space, half the Moon is always lit. Why do you see phases of the Moon on Earth?
3. What causes the seasons?
4. Why did people create stories about the constellations?
5. What is the Big Dipper and how did it get its name?
6. When our solar system was formed. What happened to the matter that did not become a star or planet?
7. What is the difference between a solar and lunar eclipse?
8. Why is there microgravity on the International Space Station?
9. What are meteors, meteoroids, comets and asteroids?
10. What happens when a meteoroid goes through our atmosphere?
11. What keeps the International Space Station from crashing into our planet?

A Little Astronomy Science Terms

1. *Asteroid*: Small planet-like bodies formed in the universe; not big enough to become a planet
2. *Asteroid Belt*: Place between Mars and Jupiter where most asteroids are found in our solar system orbiting the sun.
3. *Chicxulub Crater*: a crater formed by an asteroid and associated with the extinction of dinosaurs and most species of life on our planet about 66 million years ago.
4. *Comet*: A celestial object, made of rock, ice, dust and gases with a very elliptical orbit around Earth.
5. *Constellation:* An arbitrary formation of stars perceived as a figure or design named after characters from classical mythology and various common animals and objects.
6. *Earth*: the third planet from the sun
7. *Exoplanet*: A planet that revolves around a star outside of our solar system.
8. *First Quarter*: Moon appears as a half-moon with right side illuminated. It follows a waxing crescent.
9. *Gravity*: A property of matter that creates an attractive force that pulls matter toward other matter.
10. *Full Moon*. The entire face of the moon appears illuminated.
11. *Galaxy*: A group of billions of stars and their remnants held together by gravity
12. *International Space Station*: A orbiting space station that carries out many experiments in microgravity
13. *Jupiter:* The largest planet in our solar system
14. *Kuiper Belt*: A region of the solar system beyond the orbit of Neptune, believed to contain many comets, asteroids, and other small bodies made largely of ice.
15. *Last Quarter*: A phase of the moon whereby the left side is illuminated following a full moon.
16. *Lunar eclipse*: The moon's ability to reflect light from the sun is blocked by the shadow of the Earth as it moves between the sun and the moon.
17. *Mars*: The fourth planet from the sun and the second smallest planet after Mercury.
18. *Mercury*: Planet closest to the sun and the smallest of the inner planets.
19. *Meteoroid*: A fragment of rock formed from collisions often coming from asteroids
20. *Meteorite*: The part of a meteor that did not burn up as it entered Earth's atmosphere. Many have been collected form the South Pole because they are easy to find there.
21. *Microgravity*: A sense of weightlessness felt in a free fall.
22. *Milky Way Galaxy*: A spiral galaxy that is home to our solar system.
23. *NASA*: The National Aeronautics and Space Administration is an independent agency of the United States Federal Government responsible for the civilian space program, as well as aeronautics and aerospace research. NASA was established in 1958, succeeding the National Advisory Committee for Aeronautics.

24. *Neptune*: The outer most planet of our solar system. It is one of the gaseous planets.
25. *New Moon*: The moon reflects no sunlight because the sun is behind the moon.
26. *Planet:* An object, large enough to be round in shape that revolves around a star with or without satellites.
27. *Pluto*: A dwarf planet located in our solar system.
28. *Polaris*: The North star
29. *Saturn:* It is the sixth planet from the sun and the second largest planet in our solar system. It is one of the gaseous planets characterized by large rings around it.
30. *Shooting Star*: When a meteor travels through our atmosphere releasing light energy from friction.
31. *Solar Eclipse*: When all or part of the sun is blocked due to the moon blocking part of its view.
32. *Summer Solstice*: Occurs about March 21 when there is an equal amount of daylight and nighttime.
33. *Uranus*: It is the seventh planet from the sun. It has the third-largest planetary radius and fourth-largest planetary mass in the Solar System. It is one of the gaseous planets
34. *Venus*: The second planet from the sun and also the hottest due to it build-up of carbon dioxide.
35. *Waning Gibbous*: The phase after a full moon as the moon appears to reflect less light.
36. *Waxing Crescent*: The phase of the moon after a noon moon with a quarter of the moon appearing illuminated on the right side.
37. *Waxing Gibbous*: The phase following first quarter with three fourths of the moon illuminated on the right side
38. *Waning Crescent*: A phase right before the noon moon with one-fourth of the moon illuminated on the left side.
39. *Winter Solstice*: Occurs about December 21 when the northern hemisphere receives the least amount of sunlight so has the shortest amount of daylight.
40. *Ursa Major*: Also called the big dipper
41. *Ursa Minor*: Also called the little dipper and the tail points to the North star.
42. *Zodiac*: The zodiac is an area of the sky that extends approximately 8° north or south of the ecliptic, the apparent path of the sun across the celestial sphere over the course of the year. The paths of the Moon and visible planets are also within the belt of the zodiac.

Moon Clock Answers

PHASE	RISES	IN EASTERN SKY	HIGHEST IN SKY	SETS	IN WESTERN SKY
NEW	6:00 AM	√	NOON	6:00 PM	√
WAXING CRESCENT	9:00 AM	√	3:00 PM	9:00 PM	√
FIRST QUARTER	NOON	√	6:00 PM	MIDNIGHT	√
WAXING GIBBOUS	3:00 PM	√	9:00 PM	3:00 AM	√
FULL	6:00 PM	√	MIDNIGHT	6:00 AM	√
WANING GIBBOUS	9:00 PM	√	3:00 AM	9:00 AM	√
LAST QUARTER	MIDNIGHT	√	6:00 AM	NOON	√
WANING CRESCENT	3:00 AM	√	9:00 AM	3:00 PM	√

NGSS Articulated Standards

Disciplinary Core Idea

- Gravity holds solar system together
- Motions in the solar system
    Eclipses
- Earth is part of Milky Way Galaxy, one of millions and millions
- Solar system includes planets, comets, asteroids, moons & meteoroids
- Seasons are caused by tilt of Earth

Science and Engineering

- Using Models and analyzing and interpreting data.

Cross Cutting

Using Models, interdependence of science, engineering and technology.

Earth Shapes Up

Students attending the Martin Luther King Jr. Middle School were being dropped off at their school at 6:00 AM for a field trip to the Adirondacks in upstate New York. They were divided into groups of four led by a parent or teacher.

The past few days in their science classes, they prepped for the trip. They were taught a lot about the processes that shape their planet, but this trip was to provide them with some first-hand experience.

The buses were lined up in front of the school. The students got onto their assigned buses. All of the students brought backpacks that included their lunch, extra water, some snacks, sunblock and insect spray.

"I have never been to the Adirondacks," Alba said to her friend Marva, who was sitting next to her on the bus.

"You have never been to the Adirondacks?" exclaimed Nelson, who was sitting behind the girls. "That is hard to believe."

"Well there are people who live in New York City that have never been to the Empire State Building," Alba countered.

"Just because you live near an area does not mean you have visited it, "said Winston, who was sitting next to Nelson.

The buses soon left the school and headed to the New York State Northway. They were all going to hike Mt. Jo, a relatively simple hike, have lunch at the top and then head back down. In class, the week before, they were given some background about the geology of the Adirondacks.

"Use all of your senses while you hike," said their teacher, Mr. Anderson, "and the Adirondacks will unlock to you its amazing story."

After about 2 1/2 hours, the buses arrived at the Adirondack lodge. The students got off the buses and met up with their team leader. Marva, Alba, Winston and Nelson had signed up to be in the same group.

"How long is the trail?" asked Marva.

Their team leader, Mrs. Wilkins said, "It is about 2 1/2 miles round trip. When we get to the top, there is a beautiful view of the high peaks in the Adirondacks. We will have lunch there. After lunch, I will collect all your trash. During the hike, you must stay on the trail to avoid causing erosion around the trail"

Before, Marva, Alba, Lincoln and Winston's group began their climb, they walked over to Heart Lake.

"How do lakes form?" asked Winston.

"They form in different ways, but this one formed because of glaciers," replied Lincoln.

"You are kidding!"

"No, I am not. I read a guide about this area and it said that many of the lakes in the Adirondacks formed from glaciers. I think Mr. Anderson may have forgotten to talk about that in class last week. They are called 'kettle lakes.'" According to the guide, when glaciers move, sometimes parts break off, get buried with soil and gravel and when they melt, leave depressions. When these depressions fill with water, kettle lakes form.

"Why are they called kettle lakes?" asked Marva.

When no one answered, their team leader, Mrs. Wilkins said, "The name is derived from the old iron basins used to heat water."

"Oh, I guess I can imagine a giant pot when I look at the lake," said Marva. "Though it is a heart-shaped pot."

As the students climbed Mt. Jo, they noticed a number of creeks running down the mountain. Mrs. Wilkins asked her team if they could explain how the creeks formed.

Nelson watched the water run over the many rocks in the creek. He knew the water came from melting snow on the mountain. "When the water flows down a mountain, it weathers and erodes the mountain's surface rocks. The steeper the mountain, the faster the water travels and the more erosion it causes."

Before Nelson could finish, Marva added, "And the more water running down the mountain, the wider and deeper the creek. Of course, the water would not travel down a mountain if it was not being pulled by gravity."

"That is definitely true," agreed Mrs. Wilkins.

Alba was taking in all this new knowledge. She had seen pictures of the Grand Canyon. "So, the Grand Canyon must have been carved out by water too."

"Yes," replied Mrs. Wilkins. "We all can appreciate the powerful force of running water. The Colorado River carved out the Grand Canyon millions of years ago."

Lincoln picked up some of the rocks along the creek. He noticed how rounded the edges of the rocks were compared to rocks not in the water. "Look at the neat rocks I found."

"No wonder they are smooth. Water has been running over them, polishing their edges" said Marva.

"Another example of water causing erosion," said Lincoln.

"No, I think weathering is a better explanation because particles were broken off the rock to smooth it out. Erosion is when water transports the rock particles to a different place. Anyway, that is what I read in our science book before the trip," said Marva.

About a half mile into the hike, a light rain shower began. The students were told to bring rain jackets since the weather forecast did mention possible showers. Mrs. Wilkins brought along some extra plastic ponchos for students who forgot to bring a rain jacket. Winston gratefully accepted one of those.

"This rain is making me thirsty," said Alba as she took a drink of water.

Not even five minutes passed when the sun came out again.

"I am soaked," said Marva. "I am glad it finally stopped raining so I can dry off."

"Yup, the sun will help dry us off as the water evaporates." Alba said smiling trying to sound scientific.

Mrs. Wilkins decided to use the students' discussion about water as a teaching moment. She said to her group, "Many things on our planet get recycled. Water is one example. Who can explain what the 'water cycle' is? You all have witnessed a bit of it today."

After a minute of thinking, Winston decided to give it a try. "Most of the water on our planet is in the oceans, some in rivers, lakes and creeks and the remainder is underground or is what we call ground water. People who have wells get their water from there. The sun causes water to evaporate into the air, changing into a gas."

"So, the clouds we see is water as a gas?" asked Marva.

"No," replied Winston. "When water changes phase into a gas, it is colorless and odorless. You cannot see it."

(Credit: Howard Perlman, USGS. Public domain.)

"So, what am I seeing when I look at a cloud?" asked Marva.

"You are looking at water droplets that are not heavy enough to fall as rain," answered Winston.

"How did the cloud form?" asked Lincoln.

Mrs. Wilkins decided not to interfere in their discussion unless the students asked her a question.

"Well, when water warms up, it evaporates, but when it gets cool enough, it condenses back into a liquid. Clouds form when water condenses," Winston explained.

"It can also be a solid," added Alba. "If it gets cold enough, water freezes at 0°C." Alba

Earth Shapes Up, <u>Tales of Science</u> by Joan Wagner

was pretty familiar with the metric system since her parents were from Mexico. "In fact, if it is cold enough, some clouds would have ice crystals in them."

"That is very observant, Alba. Clouds form when ice freezes into crystals or condenses into water droplets. They are seen because light can reflect off of solids and liquids. The atmosphere contains varying amounts of water vapor, but light cannot reflect off of that"

Marva decided to summarize for the group. "So, water cycles from liquid to gas to solid and back again. And all the water we have on our planet is constant, but can exist in those three phases."

"Well, I think you all have a handle on the water cycle," said Mrs. Wilkins. Just then she told the kids to wait a minute while she went into the woods.

"I think Mrs. Wilkins is contributing to the water cycle," Winston joked.

When they began walking again, Marva yelled "ouch," as she tripped on a root and almost fell into a large boulder. "How did that get here? It doesn't look like any of the other rocks on this trail."

"Remember, the Adirondacks were eroded by glaciers. When glaciers move, they carry a lot of debris in the form of gravel and rocks. What you are looking at is what geologists call a 'glacial erratic.' It was left behind by a glacier," said Mrs. Wilkins. "It is not necessarily native to this region, "she continued.

"Hard to imagine anything could move that boulder!" Said Marva.

The last part of the hike was the steepest, but soon all of the students were on the top of Mt. Jo. As they looked out, they could see a number of the high peaks of the Adirondacks. Mrs. Wilkins told the students that there are 46 high peaks in the Adirondacks. To be classified as a high peak they have to be at an elevation of over 4000 feet though technically, a few were not that high because of errors in early measurements. She pointed to Mt. Marcy, which is the highest peak in the Adirondacks and also in the state of New York.

The students were all admiring the view and how spectacular the mountains appeared as they ate their lunches. They all found a comfy spot to sit on Mt Jo's boulders.

In their science classes, the students had learned about the Theory of Plate Tectonics. As with all scientific theories, a tremendous amount of evidence supported this theory. The students were taught how Earth's crust is divided into a number of plates that move on the layer called the mantle. As they move, they may collide, denser plates sink under less dense plates or they can slide past one another, often rubbing. The motion of plates explains mountain building also known as orogeny, volcanic eruptions and earthquakes.

Mrs. Wilkins thought it was a good time to talk about mountain formation since they were sitting on the top of one of the smaller peaks in the Adirondacks and had a nice view of some of the higher peaks.

However, before Mrs. Wilkins could lead a discussion about plate tectonics, Marva said, "I bet the Adirondacks were formed by plate tectonics."

"Glad you brought it up, Marva, because that is exactly what I wanted to discuss with all of you. There are different ways in which mountains can form

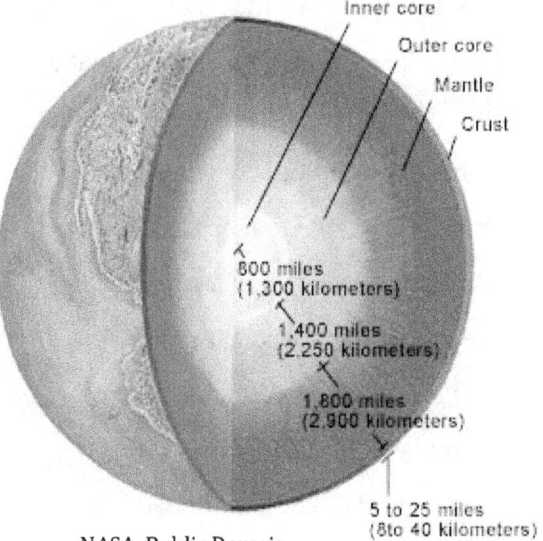
NASA, Public Domain

due to the movement of tectonic plates. However, the bottom line for all mountain building is the crust of Earth must somehow get uplifted," said Mrs. Wilkins.

"Except for volcanic mountains," noted Winston.

"Yes, we should not forget the volcanic mountains we have in the world," replied Mrs. Wilkin.

"Volcanic Mountains form when molten rock or magma pushes through an opening in the Earth's crust. As the molten rock piles up, a volcanic mountains form. Last summer, my family drove cross-country to Washington and we hiked the Cascade Mountains. They are volcanic mountains," said Winston.

"That must have been a lot of fun for your family," said Mrs. Wilkins, smiling.

"I read the Rocky Mountains are called folded mountains because they formed when tectonic plates collided and pushed the solid crust up," noted Marva

"Yes, that is true, but scientists are still studying what plate activity was responsible for the uplifting. The Rocky Mountains are a young mountain range. They are still growing," Mrs. Wilkins told them

"So how did the Adirondacks form?" asked Winston.

"They are not folded mountains like the Rockies. However, a tectonic plate was uplifted and formed what geologists called a tectonic dome mountain.

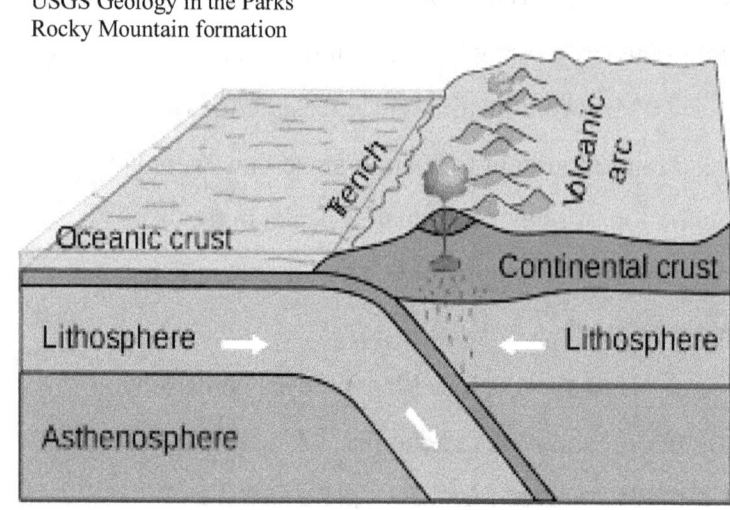

USGS Geology in the Parks
Rocky Mountain formation

If you look at this picture in my guidebook of the Adirondacks, it looks like a dome. However, what is really neat about the Adirondacks is they sit on top of an ancient mountain range that eroded away millions of years ago. According to geologists, that ancient mountain range was once as tall as the Himalayas!" stated Mrs. Wilkins.

"Then the Adirondacks must have some really old rocks if some of the ancient mountain also got uplifted," noted Lincoln.

"Yes, the Adirondacks have some of the oldest rocks in the world because of where it formed," replied Mrs. Wilkin.

Adirondacks, I, Ruhrfisch / Public domain

Marva thought about all of this and then made this observation. "So, there must be two forces involved in mountain building: erosion and uplifting. If erosion occurs at a faster rate than uplifting, then eventually the mountain will no longer exist, like the ancient mountain range on top of which the Adirondacks formed."

"That is true," replied Mrs. Wilkins, impressed with Marva's intuitive mind. "However, erosion also can shape a mountain. Though geologists have evidence the Adirondacks are still growing, it was the process of erosion that carved out the many peaks of the Adirondacks.

Since Marva's family had travelled to the Catskill Mountains in New York last summer, she asked what type of mountains they were.

"I heard they are not really mountains," said Lincoln.

Mountains are not geologically mountains because there is no folding of the bedrock. It began as a plateau that got carved into peaks by running water and glaciers."

"That is correct," said Mrs. Wilkins, again impressed with her students. "The Catskill

Mrs. Wilkins checked her watch and told her group to get ready to descend Mt. Jo. Everyone cleaned up their lunches, placing all garbage in a small bag and then into their backpacks.

"Remember what Mrs. Wilkins said, 'What we take in must go out.'" We do not want to pollute the Adirondacks with our trash," Alba stated to her friends. Alba was also excited about the climb down because she wanted to collect some of the rocks. She loved to collect rocks and minerals and study them. However, she knew she was not to remove anything so she took pictures instead.

"Can we look for the very old rocks when we hike down?" asked Alba.

"Absolutely," stated Mrs. Wilkins, pleased that the students had taken an interest in the geology of Mt. Jo.

"Always look down so you do not trip on rocks or roots," Mrs. Wilkins reminded the students. She recalled a number of times she had slipped hiking the Adirondacks, which can be tricky because of the exposed roots and slippery rocks.

As they climbed down the steepest section of Mt. Jo, Alba found a rock with a number of striations in it. "Look, my rock is striped."

Lincoln remembered looking at a rock like that in class. "That is a very nice gneiss rock," he said with a smile. "No pun intended."

Mrs. Wilkins team stopped to look at the rock.

"Why is it striped?" Marva wondered.

"Isn't that a metamorphic rock?" asked Lincoln.

"Yes," replied Mrs. Wilkins. "Most of the rocks in the Adirondacks are metamorphic."

"Metamorphic rocks are formed under extreme pressure," stated Winston. "There must have been a lot of heat and pressure in the Adirondacks to have so many metamorphic rocks."

Gneiss, Metamorphic

"Very good, Winston. That is what happens when the tectonic plates move and mountains are formed," said Mrs. Wilkins.

"Why are there striations in the rock," asked Marva?

Mrs. Wilkins told them that gneiss rocks were placed under extraordinary heat and pressure causing them to behave like clay. This caused mineral grains to recrystallize and then

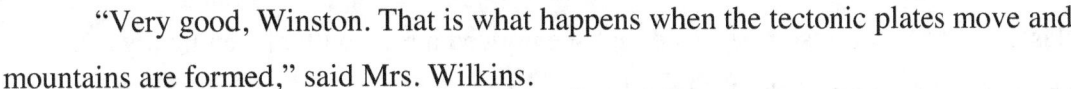

segregate into bands due differences in density, as can be seen in the gneiss rock. Marva was silent a moment and then said, "Just like the lab we did with liquid densities. The denser liquids sunk under the less dense one."

"Excellent, Marva, glad to know the labs we do in class are being applied in the field.

Lincoln picked up a rock near the creek. He noticed it was very weathered from being in the water. "This looks like sandstone, which is a type of sedimentary rock."

"That is the type of rock in which you can find fossils," said Lincoln.

"Why can't you find them in metamorphic rock?" asked Winston.

Alba and Marva listened in since they were interested in the answer.

"Isn't sedimentary rock formed from weathered rock particles that get cemented together?" said Alba.

Before Mrs. Wilkins could respond, Lincoln replied, "Yes."

Mrs. Wilkins was going to help out, but Lincoln spoke before she could.

"Easy," said Lincoln, fossils would get destroyed in rock that almost melts. The fossil would burn up in the hot rock."

"Well, so far your team has found metamorphic and sedimentary rocks. Now let's see if you can find granite. Remember, granite is considered an igneous rock that formed from magma (liquid rock) when it cools," said Mrs. Wilkins.

"Those are the rocks with the pretty quartz crystals we saw in class," said Marva. Just as she said that she noticed a rock with glistening crystals, some pinkish. She picked it up. "I think I found some granite."

"You can really see the crystals in the rock," noted Marva.

Sandstone, Sedimentary rock

Granite

"Those are the different minerals that make up the rock," said Alba. "If I remember correctly, the pink mineral is called feldspar and the opaque white one is quartz. I am not sure about the other ones. There was a sample of a rock like this in one of the labs we did in school."

"My mother has a necklace made from quartz," said Marva.

Mrs. Wilkins walked by and said, "Nice piece of granite, Marva."

"Thanks, I wish I could bring it back to the class," said Marva.

"Remember, it is important to preserve the Adirondacks. We have granite rock samples back at school," said Mrs. Wilkins.

"I wonder if granite can weather," said Alba.

"We can take some granite and expose it to running water. I can put it under the drainpipe by my house and let water run over it after a rainfall," Marva said excitedly.

"Well if it does weather, then its particles could eventually become sedimentary rock," said Winston.

Mrs. Wilkins overheard the boys talking. She said to her team of students, "All rocks can turn into other types of rocks."

"So, if sedimentary rock is placed under heat and pressure, it can become metamorphic rock" said Alba.

"Yes," replied Mrs. Wilkins. The word metamorphic means 'to change form.' Both igneous and sedimentary rocks can change into metamorphic rocks.

"All rocks can change into one another. If these rocks are pushed back into Earth's magma, they melt and when they come to the surface of Earth's crust again, they are igneous rocks," said Mrs. Wilkins.

"So, rocks go through a cycle, just like water," Winston noted.

"And the cycle is known as the 'rock cycle,'" said Mrs. Wilkins.

"But how does sedimentary or metamorphic rocks become igneous rocks?" Asked Winston.

I think I know the answer," replied Lincoln. "The tectonic plates are getting recycled too. On one side of our planet under the Atlantic Ocean is the Mid Ocean Ridge. The tectonic plates are moving apart so the Atlantic Ocean is actually getting wider." The ridge forms as magma from inside Earth moves to the surface forming an underwater mountain called the Mid Ocean Ridge.," added Mrs. Wilkins, who was walking with her team of students.

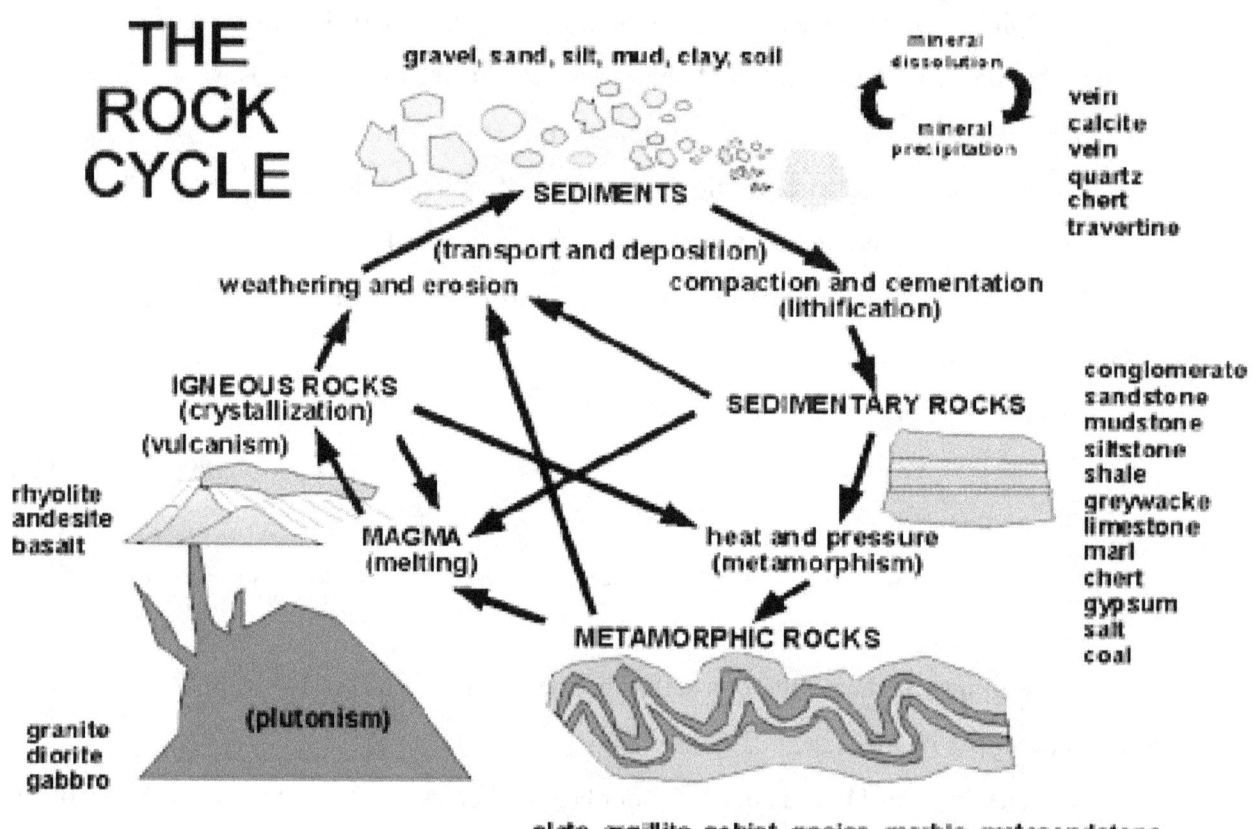

USGS Public Domain

"Well if one side of our planet is spreading apart, another side must be getting pushed together," noted Alba. "Isn't that happening in the Pacific Ocean?"

Yes, replied Mrs. Wilkins, one plate is moving under another plate in the Pacific Ocean. This plate activity is one of the reasons why that area has a lot of volcanoes and earthquakes. In fact, the region is part of the 'Ring of Fire' because of this activity."

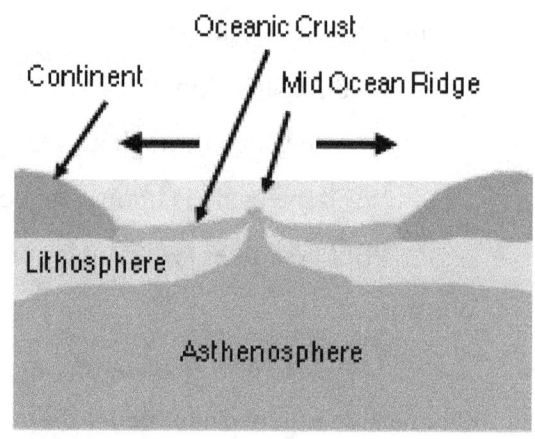

"I can see the parking lot," said Alba a little sadly.

Mrs. Wilkins checked her watch. She instructed her group of students to board the bus and prepare for the ride back to school.

"That was a lot of fun," Alba said to Marva once they were on the bus.

"We will have to try a few experiments with different rocks." said Alba.

"We can see which type of rock weathers the fastest," said Marva.

Winston and Lincoln were figuring out how they could form a model of plate

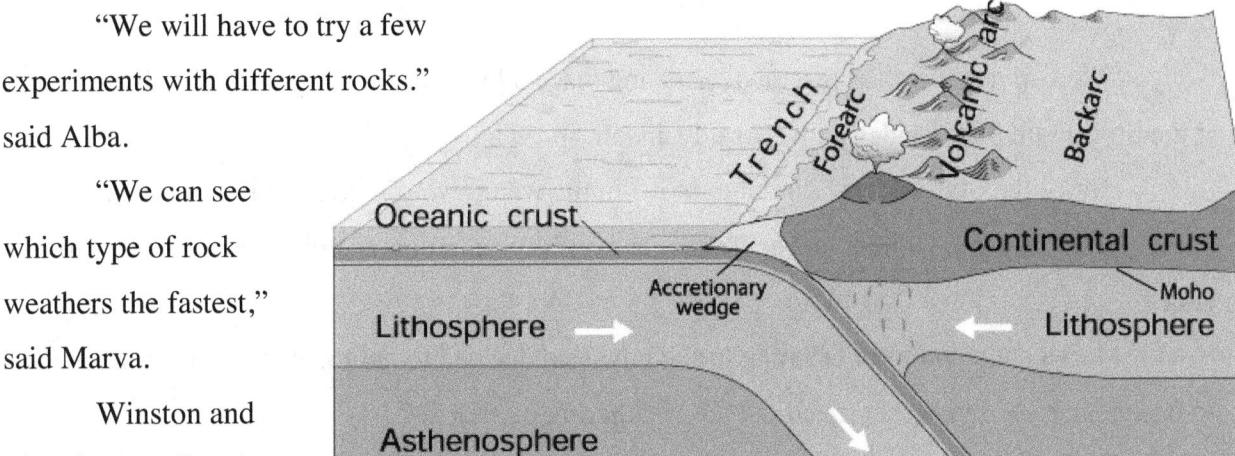

USGS Public Domain

tectonics. They thought that could be a neat project for their science class.

"Suppose we took a large glass baking pan. We can place some water in it and cut up Styrofoam to represent the plates that make up Earth's crust," said Winston.

"How do we get them to move?" Lincoln asked.

"The same way the plates move on Earth," said Winston.

"How is that?" asked Lincoln.

"Our science teacher, Mr. Howard said that since the crusts sits on hot magna, convection currents cause the crust to move," said Winston.

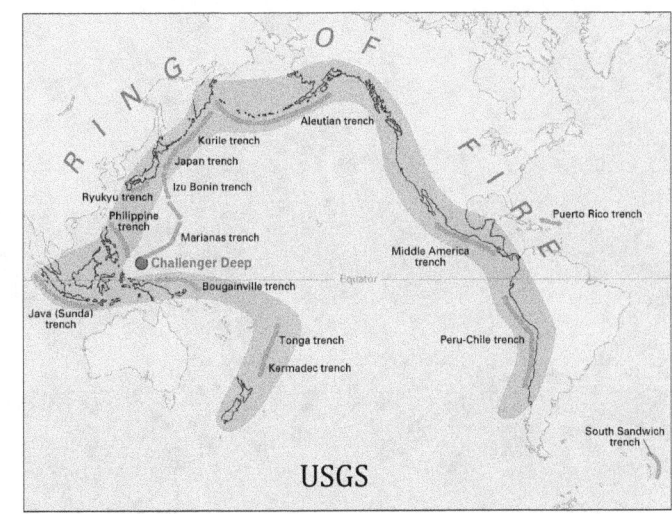

"Say what?" Said Lincoln, looking very puzzled.

"Did you ever watch a pot of water boil? The water on the bottom of the pot is closer to the burner so gets hotter faster than water near the surface. It rises to the top when the denser, cooler water on the top of the pot sinks down and pushes it up. The same thing happens in liquid rock. The cooler magna sinks down pushing up the warmer magma which pushes up against the plates, causing them to move," explained Winston. "Anyway, that is the explanation scientists use."

"We can rest the glass pan on a stand and place a lit candle under part of the dish and see if the Styrofoam plate models move," said Lincoln.

"Since the Styrofoam will tend to move in the water, we should have two set-ups, one with a candle and one without, to see if there is a difference in how the plates move," Winston added.

Marva and Alba were listening to the boys and decided to get together over the weekend to design an experiment that tested out the weathering of rocks.

On the ride back to their school, the students were pretty exhausted and some fell asleep. Winston, Lincoln, Alba and Marva had more questions about the Earth than they ever had before, but the questions made them so tired they too soon fell asleep.

Discussion Questions

1. How can a sedimentary rock change into an igneous or metamorphic rock?

2. Explain how the amount of water on Earth remains the same.

3. How are sedimentary, metamorphic and igneous rock different from one another?

4. How do mountains form (orogeny)?

5. Why is very old rock found in the Adirondacks?

6. What does plate tectonics have to do with mountain building?

7. What is the "Ring of Fire?"

8. What is the difference between weathering and erosion?

9. Why shouldn't the children take home rock samples from their hike?

10. How does a cloud form?

11. What causes the striations in gneiss rock?

12. Why isn't the Catskill Mountains considered real mountains?

13. What is a convection current and how does it explain the movement of tectonic plates?

Earth Shapes Up Science Terms

1. ***Bedrock:*** The consolidated rock under weathered rock (soil) .
2. ***Condense***: The process by which a gas changes into a liquid.
3. ***Convection Currents***: Currents formed by differences in density in a liquid or gas & molten solids. It is one of the ways by which heat is transferred.
4. ***Crust***: The outer solid layer of Earth
5. ***Crystals***: Matter with repeating, organized pattern to its structure.
6. ***Debris***: In geology, fragments of rock left over from melting glaciers
7. ***Erosion:*** The wearing down of solid earth by wind, water or other processes and its transport to other areas.
8. ***Evaporates:*** The matter left over after water is removed
9. ***Fossils***: The remains of ancient living things such as bones, shells, casts or imprints in rock
10. ***Folded Mountain***: A mountain formed when two plated collide such as the Rocky Mountains.
11. ***Geology:*** The study of solid Earth and the processes by which it changes.
12. ***Glacial Erratic***: A rock transported to another location by a glacier since it is not characteristic of the rock in that area.
13. ***Glaciers:*** A body of ice formed by the accumulation of snow. It may form on solid Earth or on the oceans. The arctic is a glacier formed over water while the Antarctic is a glacier formed over land.
14. ***Gneiss***: An example of a metamorphic rock characterized by striations.
15. ***Granite***: An example of igneous rock often associated with mountain formation.
16. ***Igneous Rock***: Rock formed from volcanic activity. It is melted rock that can cool quickly (obsidian or volcanic glass) or slowly (granite).
17. ***Kettle Lakes***: Lakes formed as glaciers carve out a depression. When they fill with water, they are called kettle lakes.
18. ***Magma***: Liquid or molten rock (semi-liquid)
19. ***Mantle***: A semi-liquid layer of Earth under the Earth's crust.
20. ***Metamorphic Rock***: Rock formed from melted rock
21. ***Orogeny***: The process of mountain building.
22. ***Plate Tectonics***: The theory that Earth is divided into a number of plates or sections that move over a semi-liquid layer of Earth.
23. ***Plateau:*** An uplifted relatively flat area of land. The Grand Canyon is a plateau carved out by the Colorado River, The Catskill Mountains in NYS is a plateau carved out by glaciers.
24 ***Quartz***: a mineral composed of silicon and oxygen
25. ***Recycle:*** The process by which something is converted to another use
26. ***Sandstone:*** An example of sedimentary rock formed from small particles of sand.
27. ***Sedimentary Rock***: Rock formed from sediments in water that stick together
28. ***Striations:*** Layers of rock formed by differences in density as the rock cooled.
29. ***Tectonic Dome Mountain***: A mountain formed by uplifting. It does not take place at a plate boundary but due to the motion of the plates placing stress on the solid earth.
30. ***Water Cycle***: The movement of water on our planet through all of its natural phases, solid, liquid and gas.
31. ***Weathering***: The chemical or physical breakdown of solid Earth

NGSS Standards addressed in this story

MS-ESS2   Earth's Systems

The History of Planet Earth
1. Tectonic processes continually generate new ocean sea floor at ridges and destroy old sea floor at trenches.

Earth's Materials and Systems
1. All Earth processes are the result of energy flowing and matter cycling within and among the planet's systems. This energy is derived from the sun and Earth's hot interior. The energy that flows and matter that cycles produce chemical and physical changes in Earth's materials and living organisms.
2. The planet's systems interact over scales that range from microscopic to global in size, and they operate over fractions of a second to billions of years. These interactions have shaped Earth's history and will determine its future.

Plate Tectonics and Large-Scale System Interactions
1. Maps of ancient land and water patterns, based on investigations of rocks and fossils, make clear how Earth's plates have moved great distances, collided, and spread apart.

The Roles of Water in Earth's Surface Processes
1. Water continually cycles among land, ocean, and atmosphere via transpiration, evaporation, condensation and crystallization, and precipitation, as well as downhill flows on land.
2. The complex patterns of the changes and the movement of water in the atmosphere, determined by winds, landforms, and ocean temperatures and currents, are major determinants of local weather patterns.
3. Global movements of water and its changes in form are propelled by sunlight and gravity.
4. Variations in density due to variations in temperature and salinity drive a global pattern of interconnected ocean currents.
5. Water's movements—both on the land and underground—cause weathering and erosion, which change the land's surface features and creates underground formations.
1. Develop and use a model to describe how unequal heating and rotation of the Earth cause patterns of atmospheric and oceanic circulation that determine regional climates.

Designing and Engineering Practices
1. Developing and Using Models
2. Planning and Carrying out investigations
3. Analyzing and interpreting data
4. Constructing explanations and designing solutions

Cross Cutting Concepts
1. Patterns
2. Cause and Effect

The Big Competition

"This can be a lot of fun," said Jameel.

"I can be the narrator," said Olive. "You know how theatrical I can get!"

"I can be the video editor," said Jameel, "I already have videos on YouTube that received many hits and excellent reviews."

"Yes! The video of your dog carrying his dog food bag and emptying it into his dish was cute," noted Olive. "Last time I checked, you had 1000 hits."

"I will volunteer to be the script writer," Olive offered.

"I wanted to do that," said Alma walking up to them at their desks.

"We can be co-authors," Olive told her friend. "It will be a big job, so it is best we work together on it."

"We have a lot of research to do," said Cheng, as he sat beside Jameel. "I will coordinate that, but first we will need to decide what information to include in our video."

The students in Mr. Novak's science class were very excited about the *Climate Change Competition*. Mr. Novak was the type of teacher who always searched for new ways to excite his students about learning. When this competition was announced online, he thought it would be both fun and educational because it was based on issues facing our planet now. Students were challenged to produce an educational video about climate change. The U.S. top 20 teams get a free trip to Washington, DC for a weekend of science activities. Each member of the team of students judged to produce the best video, would receive a Federal Bond to go toward their college education. Though the video could not be more than 5 minutes, each team also had to submit a report on climate change.

Jameel, Olive, Alma and Cheng arranged to meet in Mr. Novak's class after school. Mr. Novak said he would also give his students extra time in class some days too. When the school bell rang at 3:00 PM, the friends met in Mr. Novak's room to begin planning their video.

"We have had a really cold winter this year and some of my parents' friends think climate change is a joke. They think the weather we are having is proving that our planet is not getting warmer," said Alma.

"Well the first thing we need to do is to make sure to explain the difference between climate and weather," stated Olive.

Cheng said, "That is easy. According to what I found on the Internet, weather is what happens in the atmosphere locally, while climate is the sum of weather over a long period of time. For example, most of Arizona is dry or arid. However, even though it has a rainy season, it is considered an arid climate because the rainy season is short. On the other hand, Florida is considered sub-tropical or tropical because it gets a lot of rainfall and the temperature stays relatively warm."

"Thanks, that info should work in both our report and video," said Olive.

"I think we need to first explain why we get weather before we explain anything else," said Jameel.

"Then why do we get weather?" asked Alma.

Jameel was pretty sure he knew the answer, but decided to do a quick Internet search just to make sure. "Weather is due to the unequal heating of out planet from the Sun."

"The Moon gets unequal heating from the Sun, but it does not have weather," challenged Olive.

"Oh yeah, another criterium for weather is there must be an atmosphere," clarified Jameel. "So, weather is due to the unequal heating of our planet, which affects conditions in the atmosphere, creating high and low pressures, which in turn cause winds."

Alma added, "The different climates we have on our planet is due the amount of solar insolation it receives in a year. Because of the tilt of our planet, the equator receives the most solar insolation since the rays are most direct, while the poles receive the least amount."

"Well that explains why areas near the equator are hot and areas near the poles are cold," interjected Cheng.

"But you also have to include the amount of rainfall a region gets because that also determines its climate," added Olive.

"Oceans and other large bodies of water, in addition to the topography of the land can also impact a region's climate," added Jameel.

"I remember Mr. Novak talking about currents in the ocean and how they also can affect climate," said Alma.

"What causes currents in the ocean?' asked Olive.

Jameel checked a NOAA site and read the following to his friends: "Ocean currents are caused by differences in the density of the water. The density of water is affected by temperature and its salinity. Colder water is denser than warmer water. As the salinity of water increases, so does its density.

"Didn't Mr. Novak say the climate of the northwestern Europe is warmer than it would be for its latitude because of a current called the Gulf Stream? It brings warm water from the south. Since the temperature of the water is higher than the surrounding atmosphere, heat is transferred to the atmosphere bringing warmer air to places like the United Kingdom," said Alma, remembering a classroom discussion.

Jameel added, "Mr. Novak said new research is showing the Gulf Stream is not totally responsible for the warmer climate. The Jet Stream plays an important part too."

Before Jameel could continue, Olive interrupted. "Isn't the Jet Stream, the river of air that flows west to east around Earth and brings in the weather?"

"Yes," replied Jameel. "When the Jet Stream goes over the Rocky Mountains, it generates winds that impact the climate. The northwesterly winds bring cold air to the northeast such as New York, while the southwesterly winds bring warm air to northwestern Europe, where the United Kingdom is located. So even though the United Kingdom is at a higher latitude than New York, its climate is warmer"

"Wow, you sure know a lot, Jameel. I think my brain is going to shut down. Now we have a great deal of information about weather and climate, but we cannot have all of that in the video. Remember, the video can only be 5 minutes," reminded Olive.

"I think we are all aware of that, but we also have to write a report, so some of the information will go into the report. In the video we can just state the main difference between weather and climate and then talk about climate change," suggested Jameel.

"Predicting climate change and weather is complex," noted Cheng. "We should discuss in our report how computer modeling is used to make predictions about climate change and weather. When we listen to weather reports, it is always done with probabilities or the chance something will happen. However, they are getting better at making predictions because of faster computers and improved programs. Many of the predictions made by computer models have come out to be true such as some of the wacky weather we have been having. Hmm, maybe when I grow up I will become a computer modeling person. It sounds very interesting to me."

Mr. Novak told the friends that they should prepare to go home since the activity buses would be leaving soon. The group arranged to meet at Cheng's house that weekend. They had 3 weeks to prepare their video.

The weekend came quickly, and Alma, Olive and Jameel met at Cheng's house to work on their project.

"Let's decide what topics we want to include in our video about climate change first," started Olive.

"We should discuss the causes, evidence it is happening and what can be done about it," stated Alma.

"I was thinking," said Jameel, "that we set the video up as a panel discussion. Each of us can discuss one or two aspects of the problem."

"Who will film the video?" asked Alma.

"I can get my older brother to do it," said Cheng. "I think the video will be more interesting if we all get a chance to talk.

They worked all weekend on their project. They wrote a script about what each was to discuss and what to include in the report they had to submit for the competition. Olive's assignment was a discussion on greenhouse gases and what is meant by the *Greenhouse Effect*. Alma's assignment was evidence for climate change. Cheng would discuss how climate change is impacting weather and Jameel would discuss what can be done to slow down climate change. Their next meeting was after school in Mr. Novak's room.

Olive presented what she had researched to her friends She also found some graphics to help in her presentation. This is how she wanted to begin.

"Before we can talk about climate change, we all need to understand what is meant by the *Greenhouse Effect* and what are greenhouse gases. You know how hot the seat of your car can get if it has been sitting in the sun? That is because of the *Greenhouse Effect*. The infrared energy from the sun passes through the glass window of the car in one wavelength, but after it is absorbed by the inside of the car, it radiates back in a longer wavelength that cannot pass through the glass. Because of this, heat energy builds up in the car. "

"Makes sense so far to me," interrupted Alma.

"Please let me continue," said Olive. "We have gases in our atmosphere that sort of act like a greenhouse. Because of this, they are called greenhouse gases."

"Can you name them?" asked Jameel.

"Guys, please give me a chance to complete my presentation and then ask questions or provide me with your suggestions," said Olive, a bit annoyed. "If you take a look at these pictures, they show the major greenhouse gases and the role of human activity in their production. Before the Industrial Revolution, fossil fuels were not burned in any significant way. Since the Industrial Revolution the amount of carbon dioxide has increased by 46% and the amount of methane gas has increased by 148%!"

"Wow, that is very significant," said Cheng. "Part of the increase is also due to the increase in the world's population. In the late 1800's, the world population was a little over 1 billion. Today, it is over 7.7 billion."

"Thanks, for adding that information. Now, let me show you these pictures," continued Olive.

"This graph from the Environmental Protection Agency (EPA) shows the main greenhouse gases associated with human activity. Notice that carbon dioxide and methane are the major gases generated. Methane comes from farming practices. Hard to believe, but, cows release a lot of methane gas when they belch or fart."

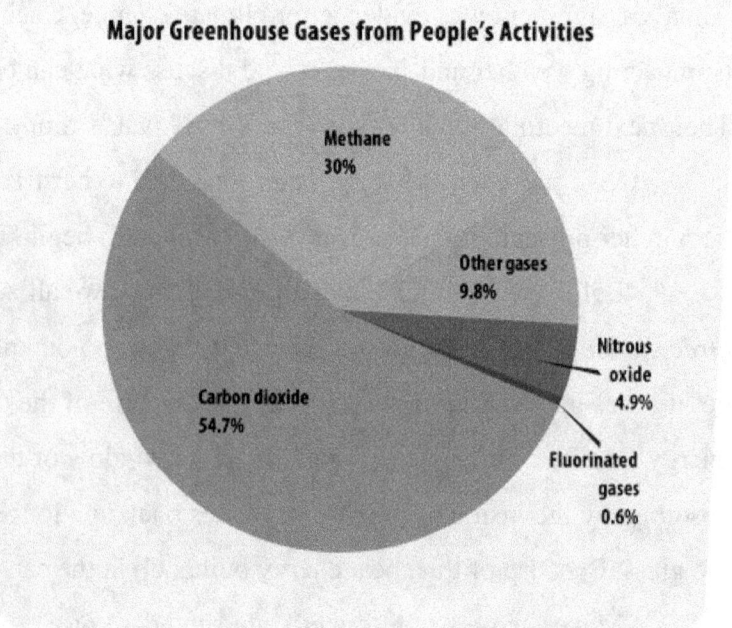

"There has to be a nicer way to say fart," said Alma, giggling, but also concerned it would offend people.

"OK, I will say 'Due to the flatulence of cows. Does that sound better, Alma?

"Yes, but I suspect some will have to look up the word."

"Carbon dioxide is released during the burning of fossil fuels," Olive continued. Some nitrous oxides come from the burning of fuels while others are produced by industry. Fluorinated gases are produced from industrial activity too.

Here is a picture that shows the main sources of greenhouse gases.

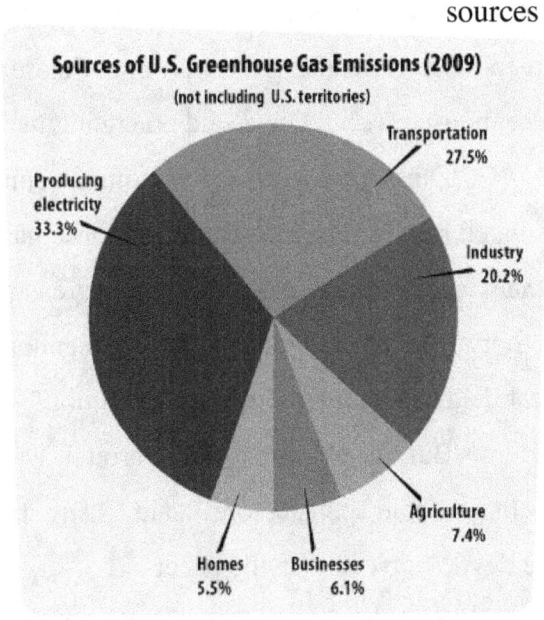

Next, Olive held up graphs showing the increase of carbon dioxide in the atmosphere since Industrial Revolution.

"These graphs also provide evidence about how the amount of carbon dioxide has increased in our atmosphere. Actually, they are not quite up to date because in May of 2013, the amount of CO2 went up to 400 parts per million. This is the first time in human history that the concentration has been that high. The last time this happened was a few million years ago when the Arctic was ice free, a savannah covered the Sahara Desert and the sea level was 40 meters higher than it is today."

CO_2 in the atmosphere and annual emissions (1750-2019)

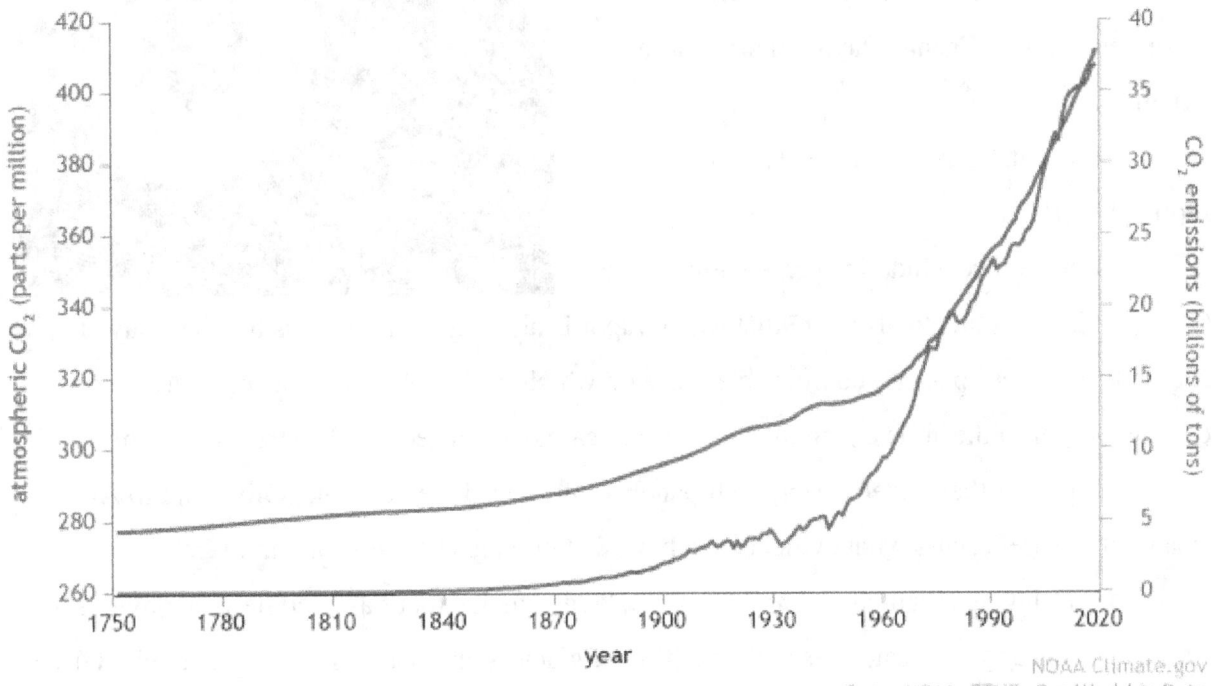

"You never explained how these greenhouse gases make our planet warmer," Cheng noted.

"OK, here is how it works. Scientists have determined that carbon dioxide is the main greenhouse gas causing the warming. Please look at the picture with the sun as I explain the greenhouse effect," Olive said practicing her presentation.

"During the day, energy from the sun warms our planet. At night some of the energy is radiated back into space. The greenhouse gases in our atmosphere have helped to make our planet a comfortable place to live. The reason our planet is getting warmer is because human activity is increasing the amount of greenhouse gases in the atmosphere so it is trapping more heat than before the Industrial Revolution."

"But our planet has gone through warm spells and cold spells before," said Cheng, playing the devil's advocate on the panel.

"Yes, that is true, but scientists have ruled factors such as changes in the tilt of Earth as causing the change in temperature," Olive countered.

"By how much has the temperature increased?" asked Cheng, playing his role as a panelist.

"It has increased by about 1.6°F or 0.9°C," replied Olive.

"Before, I conclude my presentation," said Olive, "It is important to also note that water vapor is also a greenhouse gas that has played a role in keeping our planet a comfortable place on which to live. As you will see later, the amount of water vapor in the air is changing and as it increases, it is affecting the strength of storms."

As part of their video script, each panelist introduced the next one. Olive said to Alma, "Can you please discuss what evidence we have that climate change is taking place?"

"I would be delighted to," replied Alma. "One change that has become very obvious and can be seen clearly with satellites is the melting of glaciers and permafrost. For example, Glacier

The Big Competition, <u>Tales of Science</u> by Joan S. Wagner

Park may have to be renamed. There are presently about 35 named glaciers in the Park, but only 25 are active glaciers, which means they do not keep their ice all year."

"How many glaciers existed in the Park before the Industrial Revolution?" asked Jameel.

Alma replied, "In 1850, the Park had 150 glaciers! Scientists once estimated that if current warming trends continue, there will be no glaciers left in the park by 2020!" Though this did not happen, the glaciers continue to shrink in size. Scientists are using improved models to predict the glacier shrinkage.

"Glad they all have not disappeared yet, but still not very good news," noted Olive.

"The Greenland ice sheet is melting," stated Alma. She then held up a National Aeronautical Space Agency (NASA) picture showing how much Greenland's ice sheet melted in 4 days.

"The above satellite pictures were taken on July 8 and July 12, 2012. In four days, the melt went from 40% to 97%. NASA states that this is the most satellite observed melt in the 30 years pictures have been taken."

"That is amazing," stated Olive. "And when there is less ice, less sunlight is reflected back into space, so the planet gets warmer."

"Yes," replied Alma. "As you can see, there are many factors that

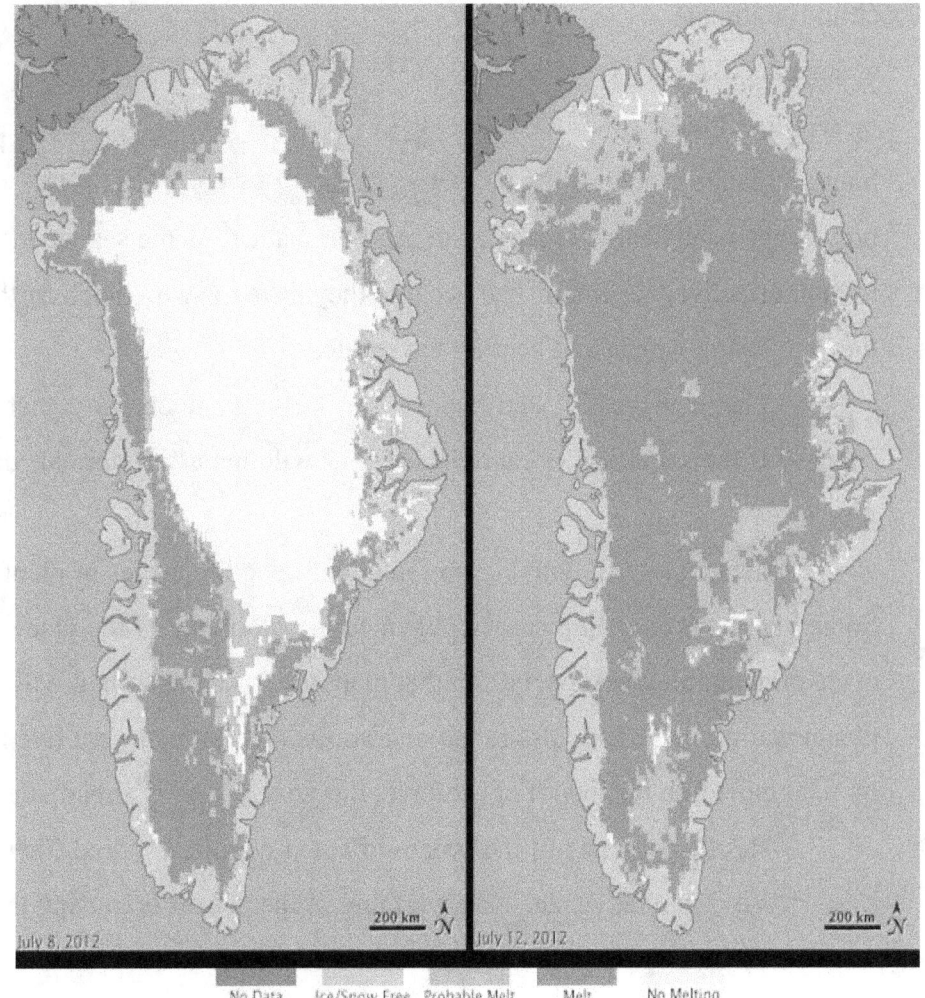

Credit: Nicolo E. DiGirolamo, SSAI/NASA GSFC, and Jesse Allen, NASA Earth Observatory

are affecting the climate of our planet and they all are connected in different ways."

"What other evidence points to climate change?" asked Cheng.

Alma replied, "Wildlife is at risk as climate change impacts their life cycles, migration and their ability to obtain food."

"I heard the polar bear is threatened," said Olive.

Alma replied, "Yes, the polar bear has become the poster child for climate change and its impact on wildlife. Polar bears hunt on sea ice. Their prey, ringed and bearded seal also depend on sea ice. As the sea ice melts, the polar bear's ability to hunt effectively is diminished because they have to swim much further to get food. As a result, they are losing weight and bearing fewer cubs.

A polar bear tests the strength of thin sea ice. Credit: Mario Hoppmann. NASA website

"That is very sad," said Jameel. "How else is life on our planet being affected by climate change? If they continue to eat reindeer, how will that affect animals that depend on reindeer for food?"

Alma continued with her presentation. "The life cycles of plants and animals are affected. Some plants are blooming earlier. When earlier blooming of the plant is not in sync with the life cycle of an animal that depends on that plant for food, the animal will be threatened. As the planet warms, some animals are moving northward. Animals that live in northern areas will have more competition for food. For example, the grizzly bear is already moving northward."

"Has the warming of the oceans affected wildlife?" asked Cheng.

"Yes," replied Alma. "The warming of the oceans is causing the bleaching of coral."

"What is the bleaching of coral?" asked Olive.

"Unicellular photosynthetic organisms such as algae live in the coral in a symbiotic relationship. The coral provides carbon dioxide for photosynthesis, while the small organisms provide food for the coral. When the oceans warm, it stresses the coral and they expel the small organisms causing them to appear white, hence the term, bleaching," replied Alma.

"If coral reefs die, then many organisms that depend on coral reefs for food and shelter will also be threatened," added Jameel.

"Yes," said Alma. In addition to that, climate change is impacting wetlands, which is home to many migrating bird species. It is also enabling the spread of invasive species."

Bleached coral NOAA picture

"What's an invasive species?" asked Cheng to ensure the audience would understand the term.

"It is a species that is not normally found in a region. For example, the Mountain Pine Beetle has caused the death of many trees in the Rocky Mountains," explained Alma. "The warmer climate has caused an overpopulation of the beetle which is interrupting the life activities of the Lodgepole pines." Continued Alma.

"Isn't climate change also causing the sea-level to rise?" asked Cheng.

"That too is another serious problem," Alma replied. "Scientists attribute sea-rise to three factors: thermal expansion, melting of glaciers and polar ice caps and ice loss from Greenland and West Antarctica."

"We should explain that thermal expansion refers to the concept that water expands or takes up more volume when it is heated," Jameel added.

"Thanks, Jameel," said Alma. "Using core ice samples, tide gauge readings, and, more recently, satellite measurements, over the past century, the Global Mean Sea Level (GMSL) has risen by 4 to 8 inches (10 to 20 centimeters). Furthermore, the annual rate of rise over the past 20 years has been 0.13 inches (3.2 millimeters) a year, which is twice the average rise of the preceding 80 years!" continued Alma.

"This will not be good news for people living in coastal regions or on islands," noted Olive.

"True, in addition to that, we also have to consider the flooding of wetlands and the contamination of aquifers by salt water," added Alma. "New research is showing how climate change is impacting weather. Cheng, will you please explain what that is all about?"

"Well, we have been having some wacky weather," Cheng started off. "According to Dr. Jennifer Francis, a scientist at the Institute of Marine and Coastal Sciences at Rutgers University, climate change is part of the reason. The warming of the planet is adding more moisture to the atmosphere, which fuels storms such as Hurricane Sandy that had a devastating impact on New York and New Jersey. Wet areas are getting wetter while dry areas, due to greater evaporation are getting drier. According to Dr. Francis, the melting of the Arctic has weakened the jet stream's westerly winds causing it to become wavier, forming large crests and troughs. When the waves are small, they move quickly, but when they are large they move more slowly. The effect is that some areas are staying very cold for long periods of time, while others are staying warm. According to Dr. Francis, a wavy jet stream was responsible for the warm weather Sochi had experienced during the 2014 Olympics."

Dr. Jennifer Francis

"Wow. So even our weather is being affected by climate change," noted Olive.

"More and more evidence is pointing to that," said Cheng.

"Is there anything we can do about climate change?" asked Alma

Since it was Jameel's role to provide ways to slow down climate change, he provided the last component of the video for the competition. Jameel found the Environmental Protection Agency's (EPA) website very helpful while doing his research.

"Since fossil fuels such as carbon dioxide play the biggest role in climate change, we should target these gases and reduce the amount we put into the atmosphere," Jameel explained.

"How can we do that?" asked Cheng.

"According to the EPA, there are four areas we can address for cutbacks. They are the home, work area, school and on the road," stated Jameel. "First, let me talk about what can be done at home. Families should use energy efficient appliances and light bulbs. LED lights and compact fluorescent bulbs are much more energy efficient. We should keep the thermostat as low as is comfortable. On cold days, wear a sweater indoors. Homes should be properly insulated to prevent heat loss. Practice the 3 R's: Reduce first, reuse second and last recycle your stuff. We need to conserve water and take shorter showers since hot water uses a lot of energy. When washing clothes, use cold water detergents. Your parents can sign up to have some of the energy delivered to their homes come from renewable sources. If you are interested, you can calculate your carbon footprint at this website: https://www.carbonfootprint.com/calculator.aspx."

"I am going to have to check out my carbon footprint," stated Olive.

"We also need to do something about how we travel," continued Jameel. "Whenever possible, we should walk, ride a bike or take public transportation because less energy is used, which translates to less carbon dioxide being dumped into the atmosphere. Our families should purchase cars that get high mileage such as hybrid cars. Electric cars would be even better."

"Taking good care of one's car is also important," noted Alma.

"Yes, it is important to keep your car tuned properly. The less bumpy the car ride, the less energy consumed."

"There are things to do when using office equipment," continued Jameel. "According to the EPA website, the total electricity consumed by idle electronics equals the annual output of 12 power plants! Most office equipment can be powered down to save energy. As at home, an office should use energy star equipment. The building should be heated and lit by Energy Star devices. Use devices with the energy star label. They are considered energy saving by the government."

"At school, we need to be better educated about climate change and how we can slow it down. We hope this presentation has been informative to students and their families," concluded Jameel.

Mr. Novak watched the final edited video presented by their group. He congratulated them on their effort and wished them luck in the competition.

Later at lunch, Alma said, "Even if we do not win anything, I learned so much doing this video plus it was also a lot of fun to produce." Her friends nodded their heads, agreeing with her.

Discussion Questions

1. Though the overall temperature of the planet is getting warmer, why do scientists prefer using the term "climate change" to "global warming," a term previously used?
2. What is meant by the "Greenhouse Effect?"
3. Explain the "Greenhouse Effect" on our planet?
4. Why is the "Greenhouse Effect" important to life on our planet?
5. Why are scientists concerned about the present warming of our planet?
6. What are "greenhouse" gases?
7. What are the main sources of "greenhouse" gases?
8. How has *Climate Change* impacted weather on our planet?
9. What can be done to slow done *Climate Change*?
10. How have living things been impacted by *Climate Change*?

The Big Competition Science Terms

1. *Active Glacier*: Glaciers that are still growing and moving.
2. *Algae:* A simple plant-like organism that manufactures food by photosynthesis.
3. *Aquifer*: An underground body of water in permeable rock that can provide drinking water.
4. *Arid*: Dry regions of the planet due to low rainfall.
5. *Atmosphere*: The gases that surround the solid earth.
6. *Carbon Dioxide*: Greenhouse gas that is most pervasive (remains in atmosphere for long periods of time) and responsible for most of the climate change. The increase of carbon dioxide in the atmosphere comes mostly from human activity.
7. *Carbon Footprint*: The degree to which carbon is used per person or object.
8. *Climate*: The average weather in a region of Earth determined by the amount of solar radiation and rainfall.
9. *Core Ice Sample*: These are drilled into the permafrost and provide a history of gas concentrations
10. *Current*: The movement of water due to factors such as wind, temperature, salinity and motion of Earth.
11. *Density:* The amount of mass per unit of volume
12. *Equator*: The imaginary circle around earth dividing it into two hemispheres
13. *Fluorinated Gases*: Human made gases that contribute to the greenhouse effect.
14. *Fossil Fuels*: Fuels that originate from ancient marine forms of life buried by sediment and placed under pressure rearranging its hydrocarbons to form present day deposits of gas and oil. Coal comes from buried plants in swamps and bogs.
15. *Glacier:* A slow moving river of ice formed from the compacting of snow.
16. *Gulf Stream*: A warm current of water originating in the Gulf of Mexico traveling northeast. It changes into the Atlantic Drift as it moves across the Atlantic to Europe.
17. *Greenhouse Effect*: The warming of earth due to greenhouse gases radiating heat back to earth.
18. *Greenhouse Gases*: Gases that trap hear such as methane, carbon dioxide, nitric oxide and water vapor.

19. ***High Pressure***: Increased air pressure on earth often to sinking cool air. The air circulates clockwise and is usually associated with dry and sunny weather.
20. ***Industrial Revolution***: The time before humans used fossil fuels (before 1800's).
21. ***Infrared Energy***: Part of the electromagnetic spectrum that we feel as heat.
22. ***Invasive Species***: Organisms that are not naturally found in a region and interfere in the life cycles of local organisms.
23. ***Jet Stream***: A current of air moving west to east and responsible for weather patterns.
24. ***Latitude***: Measurement of how north or south an area is from the equator.
25. ***Low Pressure***: Decreased air pressure on earth often due to warm air rising. The air circulates counter-clockwise. This often results in cloud formation and precipitation.
26. ***Methane Gas***: A greenhouse gas that has increased significantly due to human activities related to farming and the generation of trash.
27. ***Permafrost***: Areas of earth that stay frozen through all seasons.
28. ***Radiate***: To give off electromagnetic energy.
29. ***Rainy Season***: Part of an area that receives heavy rainfall.
30. ***Salinity***: The concentration of salts in water.
31. ***Solar Insolation***: the amount of sunlight striking earth.
32. ***Sub-Tropical***: A zone of climate characterized by hot and humid summers.
33. ***Symbiotic Relationship***: Two organisms living closely together with each benefiting.
34. ***Temperature***: A measurement of heat.
35. ***Thermal Expansion***: The expansion of matter as it is heated. Since warmer water takes up more space than cooler water. This can contribute to flooding in low regions of planet.
36. ***Tide Gauge Reading***: Used to measure tide levels.
37. ***Topography***: The layout of the land such as hills and valleys.
38. ***Tropical***: Regions of Earth near the equator.
39. ***Unicellular Photosynthetic Organism***: Organisms that produce most of the planet's oxygen.
40. ***Water Vapor***: The gas phase of water that is also a greenhouse gas.
41. ***Wavelength***: the distance from one wave crest to the next.
42. ***Weather***: The conditions of the atmosphere is a local region.
43. ***Wetland***: A region of earth that remains wet a certain amount of time each year.

Standards addressed:

Weather and Climate
1. Weather and climate are influenced by interactions involving sunlight, the ocean, the atmosphere, ice, landforms, and living things. These interactions vary with latitude, altitude, and local and regional geography, all of which can affect oceanic and atmospheric flow patterns.
2. Because these patterns are so complex, weather can only be predicted probabilistically.
3. The ocean exerts a major influence on weather and climate by absorbing energy from the sun, releasing it over time, and globally redistributing it through ocean currents.

1. Collect data to provide evidence for how the motions and complex interactions of air masses results in changes in weather conditions.
2. Develop and use a model to describe how unequal heating and rotation of the Earth cause patterns of atmospheric and oceanic circulation that determine regional climates.

MS-ESS3 Earth and Human Activity
Natural Resources
1. Humans depend on Earth's land, ocean, atmosphere, and biosphere for many different resources. Minerals, fresh water, and biosphere resources are limited, and many are not renewable or replaceable over human lifetimes. These resources are distributed unevenly around the planet as a result of past geologic processes.

Natural Hazards
1. Mapping the history of natural hazards in a region, combined with an understanding of related geologic forces can help forecast the locations and likelihoods of future events.

Human Impacts on Earth Systems
1. Human activities have significantly altered the biosphere, sometimes damaging or destroying natural habitats and causing the extinction of other species. But changes to Earth's environments can have different impacts (negative and positive) for different living things.
2. Typically, as human populations and per-capita consumption of natural resources increase, so do the negative impacts on Earth unless the activities and technologies involved are engineered otherwise.

Global Climate Change
1. Human activities, such as the release of greenhouse gases from burning fossil fuels, are major factors in the current rise in Earth's mean surface temperature (global warming). Reducing the level of climate change and reducing human vulnerability to whatever climate changes do occur depend on the understanding of climate science, engineering capabilities, and other kinds of knowledge, such as understanding of human behavior and on applying that knowledge wisely in decisions and activities.

It's Your Planet, there is No Planet B: An Earth Day Celebration

The students in Ms. Gomez's class were busy preparing their exhibits for Earth Day.

"I can't believe that Earth Day is almost here. It seems just like yesterday when Ms. Gomez told us about this project," said Griffin.

"That was in September, "replied Nikko, "And now it is April. The year certainly has flown by."

"I think it will be interesting to the spectators who visit our exhibit to learn how the energy resources of our planet are distributed," said Griffin, pleased with the research his friend Nikko and he did.

"It has been a lot of work, but I sure learned a great deal," said Nikko. "I was surprised that so many countries did not use their energy resources wisely, causing damage to the environment such as oil spills, air and water pollution," he continued.

"Yes," responded Griffin, "And in addition, they will also learn that the unequal distribution of resources results in 'have and have not' countries. Some have a lot of energy and some have little. But, even within the same country, different groups of people are fighting over energy deposits."

Griffin and Nikko prepared a map of the world that showed where the major deposits of coal, petroleum and natural gas are located. These energy resources are all considered fossil fuels. They used a map they found on a website.

"I think it is important for everyone to realize that our vital natural resources are not equally distributed across the globe," said Nikko.

"Right," said Griffin. "And my social studies teacher has shown us that the unequal distribution is one of the ways one country may try to influence another country politically. Our social studies teacher, Mrs. Dayton said this is how one country can force another country to do what they want."

Nikko agreed with Griffin, also remembering the lesson taught by Mrs. Dayton on ways in which one country can influence another. Then he thought about how large deposits of oil formed in the Middle East.

"You know, I still find it hard to believe an ocean once flowed through the Middle East. But the evidence is the large deposits of petroleum," said Nikko to his project partner and friend.

"Petroleum formed from ancient marine organisms," Nikko continued. "Layers of sediment covered these organisms, exerting enough pressure to change the remains into our modern-day deposits of petroleum and natural gas."

"Coal is also a fossil fuel, "Coal sort of formed that way too," added Griffin. "Ancient forests grew in swamps and when they died, they were buried with sediments that changed them into peat and then further burial over millions of years and pressure resulted in the formation of coal."

"We definitely should include that information in our exhibit," said Nikko.
"If you look at our map, the US, China and Russia have a lot of coal. According to the Energy Information Administration (EIA.com), about 62.7% of the electricity in our country is generated from fossil fuels (coal, natural gas, and petroleum) and 23.5% is from coal, the dirtiest of fossil fuels, but less than it used to be" continued Nikko, .

"Yeah," replied Griffin, the air is so polluted from the burning of coal in China, that many people walk around with scarves over their face to protect their lungs from all the particulates added to the air from the burning of coal. "

"Well, at least we have laws in out country to clean up the air such as the Clean Air Act, past in 1970 in response to the first Earth Day," noted Nikko.

"My grandmother said that when she used to push my father around in a carriage, she kept a nylon mesh over it to protect him, and after a short walk the mesh was covered with black particulates from the burning of garbage in New York City. However, because of the **Clean Air Act**, New York City can no longer burn its trash," said Griffin.

"Trash or solid waste is still a big problem, but I think Maria, Sasha and Nicholas will be discussing it in their exhibit, replied Nikko. "They also will be covering air and water pollution," he continued.

"In addition to explaining how these fossil fuels form, our exhibit should point out the difference between renewable and nonrenewable resources," stated Griffin.

"Definitely!" Responded Nikko. "We should explain that fossil fuels are considered non-renewable energy resources because it takes millions of years for them to form."

Nikko and Griffin both looked at their map. It was easy to see how coal, petroleum and natural gas are distributed worldwide.

"The Burning of fossil fuels also contributes to climate change because it releases carbon

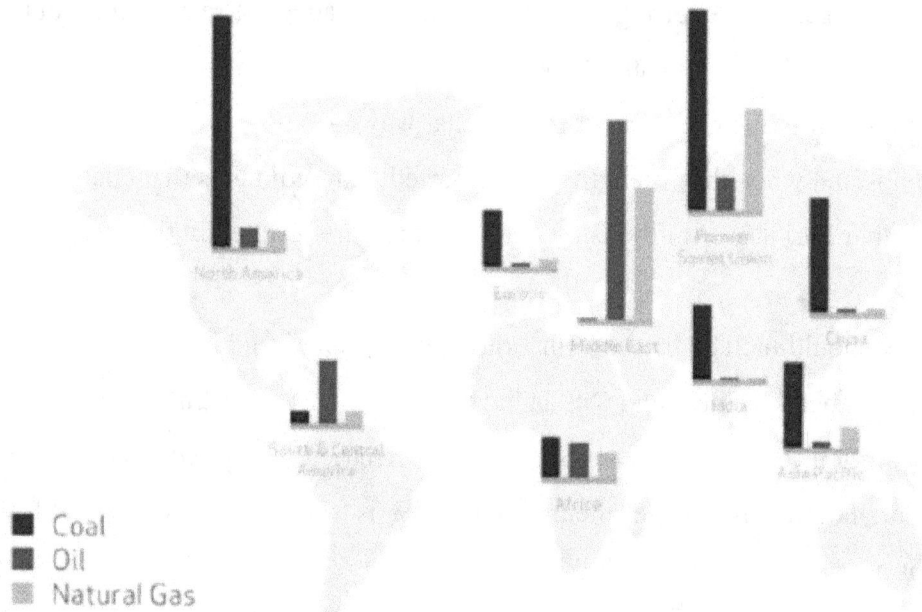

http://www.worldcoal.org/coal-society/coal-

dioxide, which is a greenhouse gas. Samantha and Ricardo's exhibit is about climate change so we do not have to talk about that in our exhibit," said Nikko.

Nikko and Griffin were putting together their posters and a diorama to show the distribution of fossil fuels. The map they found on the Internet was going to be part of the exhibit.

They also included in their exhibit a discussion and illustration of hydrofracking, a controversial way of extracting gas deposits by drilling chemically treated water into the ground horizontally. Those who are for hydrofracking say it will create jobs and natural gas and when burned releases less carbon dioxide than petroleum. Those against it state it is noisy and some of the chemicals may leak into ground water, polluting it, not to mention wasting the resource of fresh water. A lot of water is used in hydrofracking.

"My parents said our country is very divided on the issue of hydrofracking," said Nikko.

"From what I have researched the storage tanks that hold the chemically treated water are not all that safe and some of the methane seeps into the ground water," said Griffin.

"I read that some people can set fire to the water coming out of their sink because of the amount of methane in it," said Nikko.

"But I read that sometimes methane seeps into ground water naturally so what you just described can happen without fracking," replied Griffin.

"That may be true," noted Nikko, but I think our country should put its money into alternative energy.

"Me too," said Griffin, "And speaking of alternative energy, we need to decide what information we want to include in our exhibit."

"Well we should note that alternative energy is usually considered renewable energy," said Nikko.

"Wow," said Griffin, after he Googled solar energy. "Did you know that our sun, in one minute can provide the entire planet with enough energy for a year, and in a single day, it can provide the amount of energy the world population could consume in 27 years! It even says here that the amount of solar radiation that strikes our planet over a three-day period is equal to all the energy stored in fossil fuels."

"Very interesting," replied Nikko, "but from what I read, we don't yet have the technology to do all of that so we are still dependent on fossil fuels."

"My mother thinks that if the 1973 fossil fuel shortage lasted longer, we may be further ahead in the technology for alternative energy. Back then, she said, a lot of money went into alternative energy research, but as soon as the crisis ended, so did the rush to do research," stated Griffin.

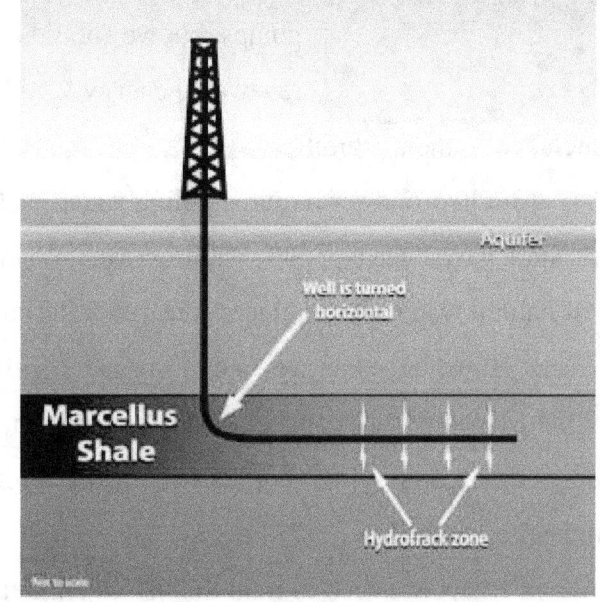

http://blogs.cas.suffolk.edu/seveilleux/2012/02/09/natural-

"She told me," continued Griffin, "the Energy Research Development Authority (ERDA) in New York State used to provide $500 grants to high school students to encourage them to do energy research. A number of her high school students won grants to compete in the ERDA competition. My father invested in ground water heat-pumps that take advantage of the fact that

It's Your Planet; There is No Planet B: An Earth Day Celebration, <u>Tales of Science</u> by Joan Wagner

the temperature 30 feet or more into the ground remains relatively stable so heat can be transferred in or out of it for the heating and cooling of a home. But, because of the initial cost of the heat pumps, that investment did not do very well. However, today, ground water heat-pumps have become more popular. Too bad my dad is not around to see his idea become a success," said Griffin. Griffin's dad had died when he was 5 years old.

"In addition to solar energy, we should also provide information in our exhibit on other types of renewable energy. For example, wind energy is getting very popular. There are even ski areas that used wind energy to run the chair lifts," like Jiminy Peak in Hancock, Massachusetts. I have skied there and it is really cool to see the big wind turbine by the slope," said Nikko.

"OK, so we have solar, wind and geothermal, such as in heat pumps, but we should also note that hydropower is a form of renewable energy," said Griffin as he reviewed what else needed to be included in their exhibit.

"But hydropower requires falling water. Natural waterfalls, such as Niagara Falls in New York State have been tapped for hydropower. When our country ran out of natural waterfalls, engineers created waterfalls, by damming up rivers. However, when we do that, we mess with 'Mother Nature.' Migrating fish cannot climb over the dam, so fish "ladders" were engineered, but many fish still fall into the turbines," noted Nikko.

"Perhaps we need to train the fish to climb the fish ladder," joked Griffin. "Yes, hydropower does have a downside, but it is still clean energy and it is renewable even though some fish get sacrificed."

"Don't forget biofuels, fuels that come from plants such as corn," added Nikko. "They are renewable and make up about 10% of the gasoline we put in cars."

"But even biofuels are controversial," noted Griffin. It uses a lot of arable land and can cause the price of produce to go up. My neighbor buys petroleum for his lawnmower that does not have alcohol." He said the alcohol damages small engines

"According to the EIA, the US generated 19.7% of its energy by nuclear in 2019. Do you think we should include nuclear energy as renewable energy because it doesn't release carbon dioxide, adding to climate change?" Asked Nikko.

"When I researched nuclear energy, it can be considered a form of alternative energy, but since the uranium gets used up, it is not renewable. In fact, the wastes from nuclear power plants are dangerous because they are still radioactive for thousands of years so the storage of the wastes is controversial," said Griffin.

"You wouldn't want to store the radioactive wastes in places that have frequent

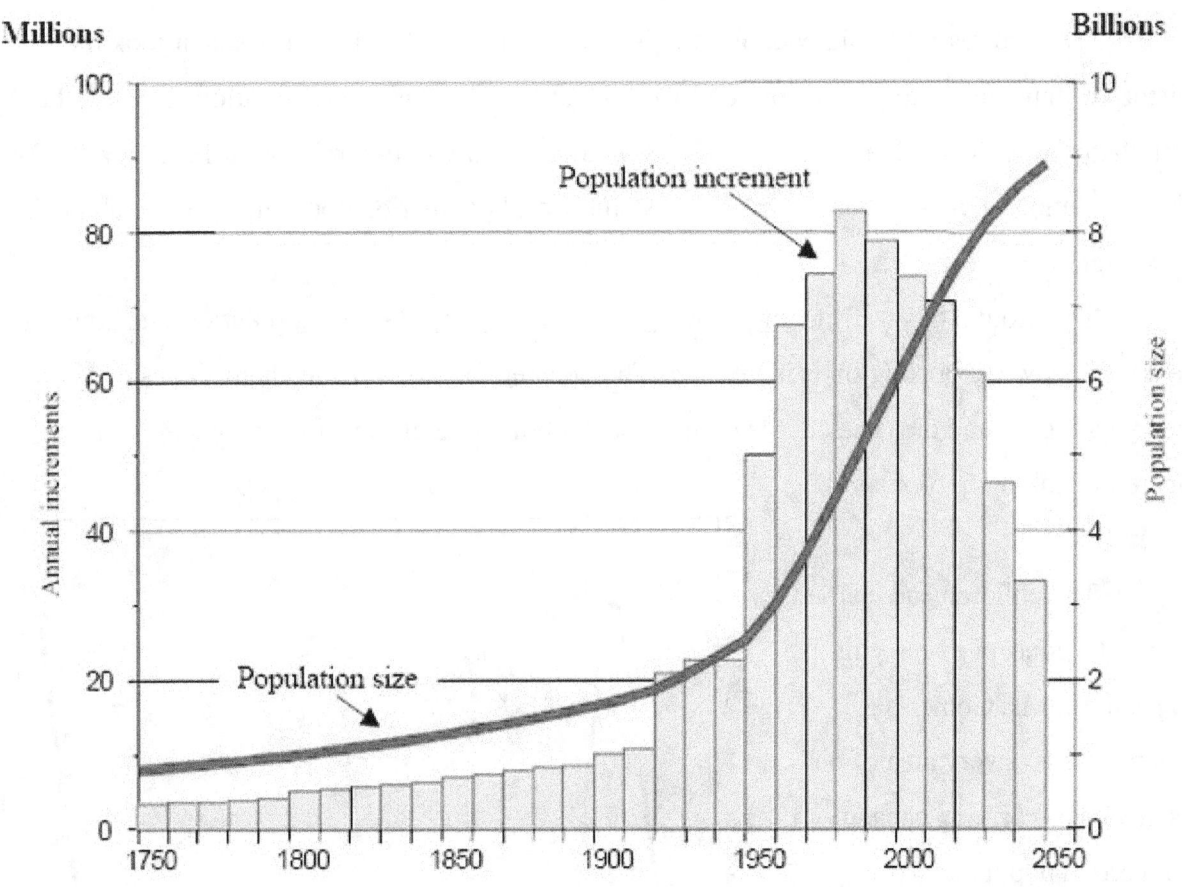

https://www.google.com/search?q=graph+of+world+population+growth+since+industrial+revolution.&clie

XNtva4BW9a1e3iwvI2uBGsuSvEw%3D&sa=X&ei=26vfU5OSJMueyASZnICwAg&ved=0CCMQ9QEwAQ&biw=

earthquakes," noted Nikko. "In fact, there is an old saying called NIMBY meaning 'Not in My Back Yard.' People would not want to live near a site where nuclear wastes are stored."

"That said, we should make sure to include the pros and cons of using all of the types of energy in out exhibit," said Griffin. "Our exhibit should help people make informed decisions about energy use," he continued.

Near Griffin and Nikko were Nicholas, Maria and Sasha working on their Earth Day exhibit. Their focus was on air, water and land pollution. They were surprised to learn of another old saying, "the solution to pollution is dilution." They realized this might have been OK when our planet had a population of less than a billion (before the Industrial Revolution in the early 1800's), but not anymore with a present world population of over 7.8 billion people.

"Do you realize," said Nicholas to his project partners, Maria and Sasha, it took the history of humankind to get to 1 billion in 1830, but as the Industrial Revolution took off, the population literally doubled to 2 billion in 100 years. So, it was two million in 1930! By 1960, the population increased by another billion so the world population doubling rate was at an all time high of 35 years. "

They found this website: http://www.worldometers.info/world-population/. It shows, in real time, how the population is increasing. The numbers on the website change constantly as people are born and die. They did learn that the doubling rate of the world population has decreased taking 61 years today.

The children had learned about the concept of carrying capacity in Ms. Gomez's science class. Every area of our planet has a carrying capacity for certain populations of organisms. When regions get overpopulated something has to give. For example, if a population gets too large, some may not be able to get enough food and will die of starvation.

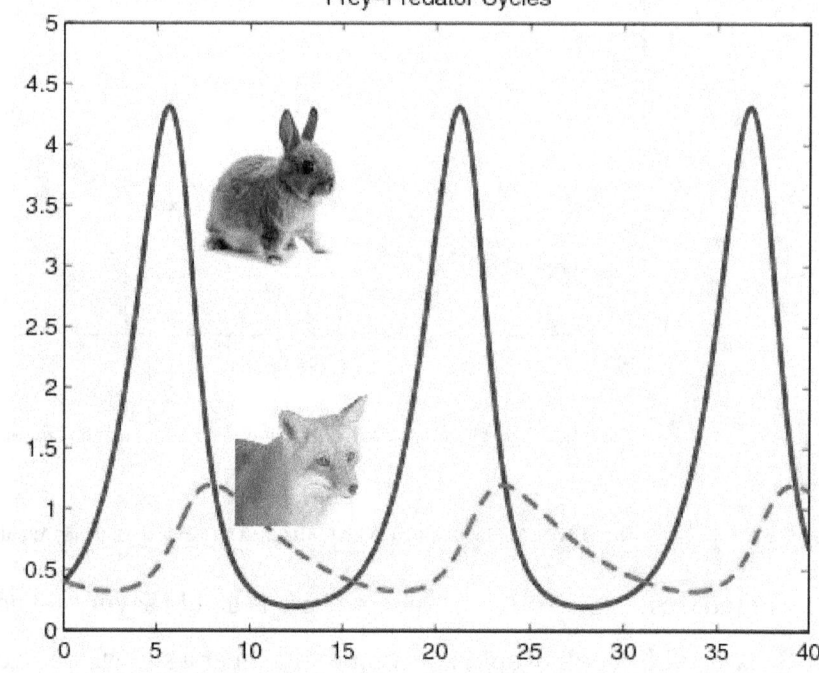

https://www.google.com/search?q=fox+and+rabbit+population&client=firefox-a&hs=S7g&rls=org.mozilla:en-

It's Your Planet; There is No Planet B: An Earth Day Celebration, **Tales of Science** by Joan Wagner

There are always checks and balances for indigenous (natural) populations. For example, when the red fox population in a region increases, there is a corresponding decrease in the rabbit population since foxes do like to eat rabbits. However, as the rabbit population decreases, so does the fox population so the rabbit population increases. The effect is that the population of these two animals remains relatively stable over time.

"We can use this rabbit/fox population graph to illustrate this concept," she continued.

"So, it is important we get across the idea that humans impact our environment, and the more people the greater the impact. When the world population was small, a little pollution had a small impact, but with more people, our planet was kind of telling us, you cannot do this anymore," stated Sasha.

Sasha, Nicholas and Maria all understood that the **_Clean Water Act_**, **_Clean Air Act_** and the management of solid wastes laws such as the first bottle bill law were all influenced by the first Earth Day in 1970 making people more aware of how we are mismanaging our planet.

"What will be a good title for our exhibit?" Asked Maria of her partners, Sasha and Nicholas.

"Hmm, let me think," replied Maria. "OK, we are trying to show the relationship between population growth and the health of our environment. I know, let's called it, 'The Population Connection.'"

"Duh, Sasha, we can't do that because that is the name of the organization that studies human impact."

"Oh yeah, I forgot," answered Sasha. "I know," she continued, "Let's call the exhibit, 'Human Impact, the Good, the Bad and the Ugly.'"

"I like the ring of that, let's go with that title. It should catch the eye of the visitors on Earth Day," said Maria.

Nicholas also agreed that was a good name.

The team's exhibit would illustrate the impact of people on air, land and water. It was back in the fall when they began this project that Maria would research air pollution, Sasha would research water pollution and Nicholas would look at land pollution.

"Eric and Zoe's exhibit will be on how humans have impacted wildlife while Jameel, Henry and Shauna will focus on plant life, particularly the destruction of rainforests.

Nicholas decided to first research what is meant by land pollution, its causes and then solutions.

"Nicholas, why don't you pretend we are visiting your exhibit and tell us all about land pollution," said Sasha.

Ms. Gomez had suggested the students practice talking about their exhibits. A number of local schools were going to make field trips to their school for Earth Day.

"There are many causes of land pollution," began Nicholas. Deforestation can result in soil erosion and land becoming a desert."

"Why do people cut down trees?" Asked Sasha, pretending to be a student from another school.

Nicholas answered, "There are a number of reasons forests are cleared. In very poor areas of the world, where most of our planet's rainforests are located, the forests are cleared to grow food. Forests are also cut down for wood, housing and industrial uses."

"What happens if we cut down too many trees?" Asked Maria

"Rainforests, are not very suitable for farming once they are chopped down."

"Why?" Interrupted Maria.

Open pit strip mining in Colstrip, Montana. Photo by P Primo

Nicholas replied, "Rainforests do not produce leaf letter because they can grow all year long. Trees in temperate zones have a growing season and lose their leaves. As the leaves decay, they enrich the soil, increasing the soil's fertility. Since rainforests cannot do this, after one growing season, the land becomes unsuitable for crops. The trees affect the climate, causing the area to become drier and drier. Scientists call this *desertification*.

"That is weird to have a rainforest turn into a desert," said Maria.

"I should also mention the roots of trees keep soil together preventing soil erosion, another type of land pollution due to deforestation," continued Nicholas.

It's Your Planet; There is No Planet B: An Earth Day Celebration, <u>Tales of Science</u> by Joan Wagner

"Land pollution also includes mining, though there are many laws in the United States that require mined land to be restored. Before there were laws, people would just dig up an area and leave large open pits. It was called strip-mining. It looked bad and was bad for the environment. Because there was a lack of vegetation, soil erosion took place.

"Isn't the trash we produce a type of land pollution?" Asked Sasha still pretending to be a visitor to their exhibit.

"Absolutely," replied Nicholas. "Our landfills are getting overcrowded and stressed. The good news is there has been a lot done to cut down on this type of pollution. Our country was becoming the "throw-away" society after World War II. As the population increased, so did the amount of trash. On Staten Island, a borough of New York City, the landfill, *Fresh Kill* was at one time the highest point in New York City. Today, it has been turned into a park which should be completed by about 2035 and will then be the largest park in New York City."

"What have we done to cut down on our trash?" Asked Maria.

"Well for one thing, through awareness that we had a major solid waster problem, a number of laws were passed. The Bottle Bill required bottles to be recycled or reused. The irony of that is before World War II, most bottles were reused," replied Nicholas.

"I get it," said Sasha, still play acting,

"Then there are the 3R's, Reduce, Reuse, Recycle," said Nicholas.

"We recycle whatever we can in our house," said Sasha as she was listening in on the discussion.

"Actually, recycling is the last thing we should do in the 3R's because our landfills are getting overcrowded. First, we need to reduce the amount of trash we generate. Companies over package their goods. We should purchase large containers instead of a lot of small ones. Not only does it save money, it reduces the trash. So, you see, recycling is the final step we should take to reduce land pollution for trash," explained Nicholas.

"We need to emphasize that in our exhibit because most people think recycling is the first thing we should do," noted Sasha.

Nicholas continued with his mock presentation to Sasha and Maria. "Soil pollution is another example of land pollution. Uses of fertilizers, pesticides and herbicides have contaminated the soil, and because of run-off, they get into the water causing pollution there too.

These toxic chemicals can get into our body through foods and vegetables grown in polluted water or soil. Some of these pollutants have been associated with diseases such as cancer.

Sasha, who had done the research on types of water pollution, gave her mock presentation to Nicholas and Maria.

"Water pollution and the wasteful use of water is a serious problem in our country. Did you know that less than 1% of fresh water on our planet is available for human consumption?" Asked Sasha.

"Less than 1%, that is hard to believe," replied Maria.

"But, it is true." replied Sasha. Most fresh water is tied up in glaciers, so not easily available for human use.

"What are some examples of water pollution?" Asked Nicholas pretending to be a visitor at the exhibit.

"When sewage gets into water, it adds nutrients that promote the growth of algae and other vegetation. This is referred to as *eutrophication*. Besides clogging filters, when these organisms die, the decay process uses a lot of oxygen from the water. The decrease of oxygen in the water can kill other aquatic life such as fish," explained Sasha.

"I have seen lakes like that. Boy they are smelly and look really yucky," said Nicholas.

"Sometimes industry adds nasty chemicals into the water. For example, many industries used polychlorinated biphenyls (PCBs) in electrical devices. These chemicals made their way into fresh water. Large amounts of PCBs were washed into the Hudson River near Fort Edward, New York. According to the Environmental Protection Agency (EPA), PCBs have a number of adverse health effects on wildlife, including cancer. There is also a cancer risk for humans that eat contaminated fish. Today PCBs are banned in the US. The Hudson River is being cleaned up of its PCBs," said Sasha.

"I like eating fish," said Nicholas, "but, I guess people ought to check from where the fish they eat were caught."

"What other types of water pollution are there?" Asked Maria helping Sasha practice her presentation.

Sasha replied: "Another type of water pollution comes from acid rain. According to the EPA, about 2/3 of all sulfur dioxide and nitrogen oxides come from the generation of electric power that uses fossils fuels, like coal. When these gases are released into the atmosphere and combine with oxygen and water droplets in the air they form sulfuric and nitric acids so the precipitation becomes acidic."

"I heard that hundreds of the lakes in the Adirondack Mountains of New York State had become acidic because of the burning of coal in the Midwest," noted Nicholas.

"That is because the jet stream moves west to east and carries the acidic precipitation to the northeast. In the winter, it literally snows sulfuric acid in affected areas," said Maria.

"Yes, I am aware of all that," said Sasha." Power plants in the Midwest have placed scrubbers in the power plants to cut down on the release of those gases. The result is that acid rain is not as bad as it used to be back in the 70's. Earth Day helped to bring awareness of this problem," Sasha continued.

"Are you going to talk about why the oceans are getting acid?" Asked Nicholas of Sasha.

"I was not going to, but now that you mention it, I should state that the burning of fossil fuels releases carbon dioxide, a greenhouse gas, and besides its impact on climate change, when dissolved in water, it forms a weak acid called carbonic acid. The extra carbon dioxide is causing the oceans to become acidic and interfering with the shell growth of marine animals," Said Sasha.

"I guess my dad won't be eating too many clams if that continues," joked Maria.

Maria's job was to research air pollution. Her exhibit partners next questioned her about air pollution. Maria decided to provide everyone with the definition of air pollution she found on Wikipedia. She read to her friends the following: "Air pollution is the introduction into the atmosphere of chemicals, particulates, or biological materials that cause discomfort, disease, or death to humans, damage other living organisms such as food crops, or damage the natural environment or built environment."

"Wow, that sounds like an all-encompassing definition," said Sasha to Maria, "but can you give some examples of air pollutants and why they are a problem so it makes more sense to me?"

"Definitely," said Maria. "That will be the main part of the air pollution exhibit I am putting together."

Maria used the EPA website to help her learn about the major air pollutants.

First, she reported on the gas ozone. She said, "Ozone is a gas made up of three atoms of oxygen chemically combined. Its chemical formula is O_3. It is found naturally in the upper atmosphere and protects us from ultra violet radiation, you know, the type that causes suntan. However, ozone is also formed in the lower atmosphere or troposphere."

"I gather the ozone formed in the troposphere is the bad ozone," stated Nicholas. "But how is it formed? He asked to help Maria practice her presentation.

Maria replied: "The exhaust from vehicles contains nitric oxide. In the presence of sunlight, a series of chemical reactions take place resulting in ozone formation."

"What health problems result from breathing in ozone?" Asked Sasha.

"People who have respiratory problems, such as asthma, are particularly sensitive to ozone. Even healthy people can experience irritations to their respiratory track," replied Maria.

"What can we do to cut back on this air pollutant?" Asked Sasha.

"There are a number of things you can do, but two of the most effective ways is to drive less, car pool and change to electric cars" answered Maria.

'What about using public transportation," added Sasha.

"That too," said Maria, "when it is available. In fact, walking and biking are great ways to get around. I know in some cities they have bikes that can be rented to get around the city. There are many places throughout the city to pick up or drop off bikes."

"Another source of air pollution are particulates or tiny particles of matter that get emitted into the air, usually formed when something is burning or they can be small drops of liquid," Maria continued.

"I gather those particulates will not be great for your lungs too," noted Nicholas.

"You bet," answered Maria. "Fortunately, the *Clean Air Act* has legislation that controls the release of the particulates." Maria then showed her friends a graph she got off the EPA website that showed how there has been a decrease in particulates since 1990.

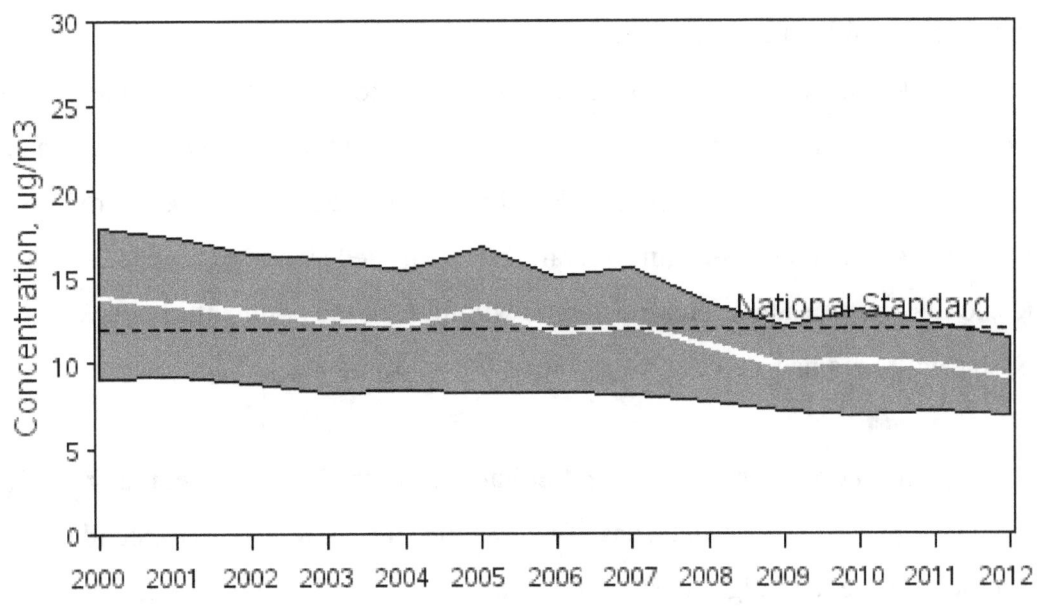

"But there are countries such as China that release so many particulates in the air from the burning of coal that the region looks dark during the day. My uncle just got back from Beijing and it was pretty bad there," said Maria. "He said many people wear masks to protect their lungs."

"Are there any more air pollutants you will be discussing in your exhibit?" Asked Nicholas.

"Yes," replied Maria. "Carbon monoxide forms when there is incomplete combustion. It can come from vehicle exhaust, fireplaces, woodstoves and even gas stoves. The gas prevents oxygen from reaching your cells so you can die from suffocation if too much of the gas is breathed in. Fortunately, most states in our country require carbon monoxide detectors along with smoke detectors."

"I have one in my house," said Sasha.

"Me too," said Nicholas.

"Are there any other air pollutants we should be aware of?" Asked Nicholas.

Maria answered, "The air pollutants of sulfur and nitrogen oxides were already mentioned as contributing to the problem of acid rain. However, since we breathe in these gases, they too can irritate our respiratory system. Those who already have respiratory disorders such as asthma, are particularly affected."

"I heard lead could pollute the air, said Sasha.

"Actually, I learned in my research we have a success story with the air pollutant lead. It used to get into the air from the burning of gasoline, but now cars use lead free gas. We can thank the *Clean Air Act* for getting that passed. But there is still lead being released from the processing of metal ores and piston-engine aircraft that use leaded aviation gasoline."

Eric and Zoe practiced their presentation on wildlife. All the pollutants discussed by Nicholas, Sasha and Maria had also caused problems for wildlife, which they explained in their exhibit.

"Do you realize how much solid wastes get into the ocean?" Asked Zoe, rhetorically, while practicing her presentation. "Autopsies of marine life such as whales have found large amounts of plastics inside their digestive tract that was responsible for their death," she continued.

"That is terrible," said Eric trying to imagine plastics inside an animal's stomach.

"The solid waste dumped into our oceans is a real problem that needs to be addressed," said Zoe.

Left: Animals can become entangled in marine debris, particularly in items such as derelict fishing lines and prevented from reaching the surface to breathe. Image credit: NOAA.

Ms. Gomez told her class that the pollution of the oceans is a global problem from which countries need to work together to solve.

Zoe practiced her presentation with Eric.

"In addition to polluted water, land and air, the building of factories, office buildings, malls, homes and roads results in less land available for wildlife. However, smart developments provide "wildlife areas or parks, said Zoe"

Eric decided to talk about the problems of invasive species. "These are organisms that get into environments that are not natural for them. Once there, they threaten the survival of the organisms normally found there," he said.

"What are some examples?" Asked Zoe, helping Eric to practice.

"Gypsy moths were brought to the United States from Asia in 1869. They were going to be used for a silkworm industry. Some escaped captivity and soon became a major pest in the northeastern United States and southeastern Canada destroying oak trees. I read that some affected areas looked like fall in the summer because of the lack of leaves on the trees. The caterpillars were voracious eaters. My father said they were in his backyard when he was little. He said you could hear them chewing. His white house looked black because caterpillars covered the house," said Eric.

"Another example of an invasive species is the zebra mussel which came from Europe. They thrive in fresh water and can clog up pipes in the water and cause a decline in algae that other aquatic life need," continued Eric.

"Those are good examples of invasive species," said Zoe. "Can plants be invasive too?

"Yes," replied Eric. Purple Loosestrife is a flowering plant that came from Europe. In the northeast, it is growing rapidly displacing indigenous plants.

"Why does it do that?" Asked Zoe, helping Eric with his presentation.

"It thrives in wetlands, which the northeast has many and reproduces more rapidly than the local plants crowding them out. Since some of those plants provide food and cover for wildlife, animal populations are affected," explained Eric.

"Climate change is also impacting wildlife around the world. We should show pictures of how the warming of the oceans is causing the bleaching of coral reefs. Many species are getting out of sync with the food they eat because the growing season starts too early, said Eric.

"Well José and Samantha will be discussing a lot of that in their exhibit so we do not have to cover it, said Zoe.

All of Ms. Gomez's students felt pretty prepared for their Earth Day presentation after practicing all day. They were looking forward to the big event. Mrs. Gomez decided to make it a competition. She organized some of the teachers and administrators to be judges to pick out the top three exhibits. Ms. Gomez told her class that she felt every school should host an Earth Day. It was her hope that all of her students will have a better appreciation of their planet and what needs to be done to keep it a healthy place to live. Personally, she had great concern about the health of our planet. She hoped this experience would encourage her students to become stewards of the environment.

"Remember," Mrs. Gomez said to her class the day before Earth Day, "humans can have both a positive and negative impact on our planet. It all depends on the decisions you make and the more you know, the better decision maker you will become. Good luck to all of you tomorrow. I hope this experience is educational to both you and your visitors."

Discussion

1. Why do you think Ms. Gomez had her students develop projects for Earth Day?

2. What problems facing our planet concern you the most? Explain.

3. What are some of the solutions proposed to the problems facing our planet?

4. What does the saying, "There is no planet B mean?

5. What is air pollution?

6. What is land pollution?

7. What is water pollution?

8. Explain what invasive species are and why are they a problem?

9. What are renewable sources of energy and why are they considered renewable?

10. What effect does population growth have on our planet?

11. What do you think is the biggest environmental problem facing our planet today and why do you think it is a problem?

Science Terms for It's Your Plant

1. **3 R's**: This refers to the management of solid wastes. First is to reduce (R), second is to reuse (R) and last is to recycle (R).

2. **Acid Rain**: This is when rain becomes acidified due mostly from the burning of coal used to generate electricity. When coal burns, it also releases sulfur dioxide which combines with moisture in the air to form acid rain

3. **Air particulates**: Any solid material released into the air from. The incineration of trash releases solid particles.

4. **Alternative**: Energy from sources other than fossil fuels. Examples are wind, solar energy, geothermal. And nuclear

5. **Biofuels:** These are fuels made from living things such as alcohol from corn.

6. **Carrying Capacity**: Every area can only maintain a certain population of an organism in order to provide it enough food and space.

7. **Climate Change**: The climate of the planet is changing with the overall result in the planet getting warmer. It is also characterized by more intense storms and weather patterns.

8. **Deforestation**: The destruction of major forests, particularly the forests in tropical regions such as the Amazon.

9. **Desertification:** When forests are destroyed and swamps drained, areas once rich in rainfall receive less, changing into a desert.

10. **Eutrophication:** When wastes from sewage and farming practices (fertilizers) enter bodies of water, excess growth of algae and other photosynthetic organisms occurs.

When they die and decay, they removed dissolved oxygen from the water causing the death of other organisms.

11. **Fossil Fuels**: These are fuels derived from ancient life, mostly ferns (coal) and microscopic marine organisms (oil and natural gas).

12. **Geothermal Energy**: This is energy derived from the center of earth.

13. **Greenhouse Gas:** Any gas such as carbon dioxide and methane that contributes to climate change

14. **Hydrofracking**: This is the process of extracting natural gas from horizontal mining.

15. **Hydropower**: This is energy generated from the motion of water such as a water fall or tidal motion.

16. **Indigenous plants**: These are plants that are naturally found in a region.

17. **Invasive Species**: This is a foreign species that prevents indigenous species from growing, usually because of superior reproduction.

18. **Jet Stream**: This is a major air current moving across the United States from west to east

19. **Land Pollution**: This is when the land is contaminated by solid wastes, liquids and gases.

20. **Landfills**: This is a way to manage solid wastes

21. **Non-renewable Energy**: This refers to energy generated by fossil fuels. It is non-renewable since it takes millions of years for the fossil fuels to form.

22. **Nuclear Energy**: This is energy that uses radioactive substances such as uranium to generate electricity.

23. **PCBs**: These are polychlorinated biphenyls once used in capacitors, but has been linked to casing cancer and other defects.

24. **Rainforest**: These are forests found in tropical regions. They are very important to Earth. For example, 20% of the planet's oxygen is formed from the Amazon rain forest.

25. **Renewable Energy**: Energy that will not run out such as solar, wind and geothermal.

26. **Soil Erosion**: The washing away of soil often caused by the loss of plants such as trees whose roots bind the soil.

27. **Solar Energy**: Energy from the sun that can directly produce electricity with photovoltaics and heat water with solar collectors.

28. **Troposphere**: The part of Earth's atmosphere is which weather occurs.

29. **Water Pollution**: Water that is contaminated with substances that impact life negatively.

Human Impact Standards

MS-ESS3 Earth and Human Activity

MS-ESS3-1. Construct a scientific explanation based on evidence for how the uneven distributions of Earth's mineral, energy, and groundwater resources are the result of past and current geoscience processes. [Clarification Statement: Emphasis is on how these resources are limited and typically non-renewable, and how their distributions are significantly changing as a result of removal by humans. Examples of uneven distributions of resources as a result of past processes include but are not limited to petroleum (locations of the burial of organic marine sediments and subsequent geologic traps), metal ores (locations of past volcanic and hydrothermal activity associated with subduction zones), and soil (locations of active weathering and/or deposition of rock).]

MS-ESS3-4. Construct an argument supported by evidence for how increases in human population and per-capita consumption of natural resources impact Earth's systems. [Clarification Statement: Examples of evidence include grade-appropriate databases on human populations and the rates of consumption of food and natural resources (such as freshwater, mineral, and energy). Examples of impacts can include changes to the appearance, composition, and structure of Earth's systems as well as the rates at which they change. The consequences of increases in human populations and consumption of natural resources are described by science, but science does not make the decisions for the actions society takes.]

ESS3.A: Natural Resources

Humans depend on Earth's land, ocean, atmosphere, and biosphere for many different resources. Minerals, fresh water, and biosphere resources are limited, and many are not renewable or replaceable over human lifetimes. These resources are distributed unevenly around the planet as a result of past geologic processes.

ESS3.C: Human Impacts on Earth Systems

Human activities have significantly altered the biosphere, sometimes damaging or destroying natural habitats and causing the extinction of other species. But changes to Earth's environments can have different impacts (negative and positive) for different living things.

Typically, as human populations and per-capita consumption of natural resources increase, so do the negative impacts on Earth unless the activities and technologies involved are engineered otherwise.

ESS3.D: Global Climate Change

Human activities, such as the release of greenhouse gases from burning fossil fuels, are major factors in the current rise in Earth's mean surface temperature (global warming). Reducing the level of climate change and reducing human vulnerability to whatever climate changes do occur depend on the understanding of climate science, engineering capabilities, and other kinds of knowledge, such as understanding of human behavior and on applying that knowledge wisely in decisions and activities.

Science and Engineering Practices
- Asking questions and defining problems
- Analyzing and interpreting data
- Constructing explanations and designing solutions
- Engaging in argument from evidence

Crosscutting Concepts
- Patterns
- Cause and Effect
- Stability and change
- Influence of science, engineering and technology on society and the natural world

Farrah's Electric Birthday Party

Farrah was very excited. Today was her 12th birthday and she was having a party at her house to celebrate. She and her brother Eli were helping their mom decorate before the festivities later that afternoon.

"Let's blow up these balloons and stick them on the wall," their mom, Lisa, stated as she walked out of the kitchen where her husband, JD was taking out cupcakes from the oven.

"Where's the tape?" asked Farrah as she looked for the bag of multi-colored balloons.

"We don't need tape, Farrah," Lisa replied.

Farrah looked at her mom, confused. "How can it stick to the wall then?"

"Just watch," Lisa said. She blew up a balloon, rubbed it on her hair and then stuck it to the wall.

"Wow! That's cool!" the kids replied in unison.

"Do you have to rub the balloon on your hair for it to stick?" What if I rubbed it on the table? Eli wondered out loud.

"Why don't you two see for yourselves," Lisa said.

The siblings blew up balloons, tied them, rubbed them on their hair and tried to get them to stick to the wall. Eli rubbed one balloon on the table and then placed it against the wall. The balloon stuck for a second, but soon fell off so he used his hair and it worked better.

"Cool, all the balloons sticking to the wall look festive." Said Farah.

"Rats," said Farrah as she watched one of the balloons fall to the ground. "Why didn't the balloon stick to the wall anymore?"

"Think gravity." Said Farrah's mom.

"I'm thinking mom, but nothing is giving me the answer.

"Wait, I think I get it. The force of gravity must have been stronger than whatever force causes the balloons to stick to the wall."

"You got it, Einstein sister." Said her brother.

"But what forces causes a balloon to stick to the wall?

Eli loved science and thought about what he learned in his 8th grade science class. "It has to do with the makeup of matter. All matter is composed of atoms."

"Oh yeah, and atoms have positive and negative charges in them."

"Right sis, and they also have neutral charges. The positive charges are called protons and the neutral charges are called neutrons."

"And neutrons and protons are located in the center of the atom called the nucleus."

"Right again, sis. And the electrons are located on the outside of the nucleus. Since the number of electrons and protons are equal, atoms are neutral or have no charge."

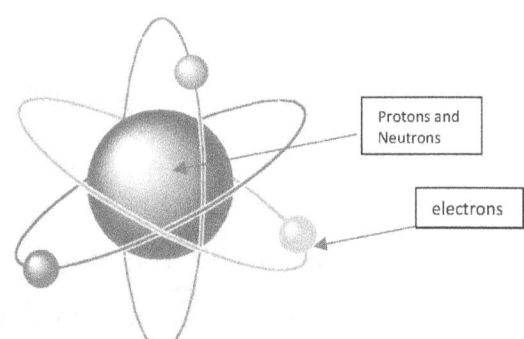

"So why did the balloon stick to the wall after mon rubbed it on her hair?' asked Farrah.

"When the balloon is rubbed on something, it is possible that some of the negative charges rub off or on the balloon." Eli said.

"Well if the balloon gains negatively charged electrons, there will be more negative charges in the balloon, giving it a negative charge."

Lisa was listening to her children and decided to see if they could figure it out on their own.

"I still don't get why the balloon sticks to the wall." Said Farrah.

Eli explained. "A science law can explain that. My teacher told us the 'Law of Charges' says that like charges repel one another and opposite charges attract. So, positive charges repel each other, and positive and negative charges will attract each other."

"Okay, Mr. Textbook, but, why does the balloon stick to the wall then?" Farrah challenged her brother.

"Since the balloon may have a negative charge, when it is is placed by the wall, it is attracted to the opposite charges on the wall. This makes it stick," her brother concluded."

Farrah let this sink in for a moment as she blew up another balloon. After she tied it, she asked, "Why does the balloon eventually fall off?"

Eli picked up the balloon his sister had placed on the table and rubbed it on his head as he thoughtfully replied: "I guess it is because it loses its charge after a while and good old gravity pulls it to the floor as you noted before."

"Close enough, though the balloon could still have a little charge left, but the pull of gravity became stronger than the attraction to the wall," their mom said. She had listened to their conversation as she helped her husband remove the cupcakes from the pan.

"Let's finish these balloons and ice the cupcakes before your friends come over, Farrah," Lisa said.

Around 4'o'clock, Farrah's friends had all arrived to her party. They sat around the kitchen table, talking about school, sports and ideas they had for their science projects, which were due next month.

Of course, they may have discussed other school subjects, but you are reading a science story!

Lisa, who was listening, planned a number of activities for them. All birthday parties her children attended these days always seem to have a theme. She decided to create, with Farrah's help, a science theme birthday party on electricity and magnetism.

"I have a few fun activities you all could do now," she said. She gathered the supplies she prepared: drinking straw, clay and lifesaver-shaped magnets. She gave each child a drinking straw, a bit of clay and three magnets. She told them to play around with the magnets for a bit and then to tell her what they observed.

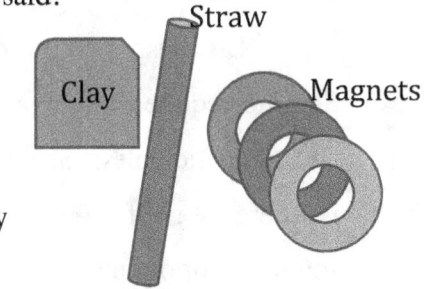

She went back to the kitchen to fill cups of punch and watched the kids as well. Some placed the straw into the clay at an angle and then added magnets. Others placed the straw vertically in the clay and added the magnets. A few first played with the magnets and she could see that they noticed some magnets stuck to one another while others repelled or pushed away from one the other.

This got Noe, a classmate of Farrah's, thinking about what would happen if two magnets stuck together were placed near a single magnet.

"Look!" she yelled out, "two magnets push harder than one magnet." She repeated the action, but placed the two magnets stuck together near the other side of the magnet. The magnets pushed away from the magnet with a force greater than when she did it with a single magnet.

"Hmm," she thought. "Two magnets stuck together can exert a greater force than one magnet."

"Benna said, "Well it looks like one side of a magnet attracts another magnet and one side actually pushes the magnet away."

Farrah thought about the balloons and how they stuck to the wall after being rubbed.

"Mom, isn't this like the charges in balloons?"

"Yes," said her mom to all the children, "but your observation is not due to the charges on the magnets, but due to its poles. Magnets have north and south poles so just like the "Law of Charges," when like poles face one another they push away or repel one another, while unlike poles attract one another. This is called the 'Law of Poles'"

"So, the north and south poles of Earth are magnetic?" Asked Carlos.

Lisa with a twinkle in her eyes said, "Yes, but we have two sets of poles."

Carlos and the other children looked very puzzled. Eli was trying to think about his science class but was going blank.

Lisa suspected this question may come up and was prepared to provide an answer. She showed the children a picture of Earth with its geographic and magnetic poles. "See, our geographic north and south poles is where we go if we wanted to reach the most north and most south part of our planet. Notice the magnetic poles are not located there."

"You mean, Earth is a giant magnet," said Carlos.

"Exactly," said Lisa. "In fact, some scientists believe some animals use the magnetic poles to navigate Earth.

"Now that is cool," said Carlos.

As she was explaining this, Eli was carefully placing his straw in the clay so that it was perpendicular to the table where he was sitting. He then placed the magnets on the straw.

"Hey, look at this. Some of my magnets float on the straw!" he exclaimed.

"That's neat," Noe said.

Eli, remembering that magnets can attract or repel, realized that the floating magnets were simply repelling one another on the straw.

"I must have placed the magnets on the straw so that like poles were pushing against one another, creating the spaces between the magnets," he observed. One of the things he also noticed was that the spaces between the two bottom magnets got smaller when he added the third magnet. The other party guests followed Eli's lead and had the same results as he.

This puzzled Eli. None of the other children seemed to know why this happened either.

Noticing their confusion, Lisa asked them, "Can you name the force that causes the magnet to slide down the straw?"

"Gravity," answered Allison, one of Farrah's friends.

Lisa nodded. "Correct. When the magnets are placed on the straw, even though they repel, they are also being pulled down at the same time. Magnets near the bottom of the straw have more weight pushing down on them. This causes the space between the magnets near the bottom of the straw to get smaller."

They all nodded, understanding this explanation.

"Look, the magnet is attracted to the clip in my hair," said Noe.

"Is your hair clip made of metal," asked Eli.

"It is made of metal and plastic," replied Noe.

Farrah's mom was listening in on the conversation and decided to add a mini experiment. She placed on the table an assortment of different types of matter, rubber band, paper clips, cardboard, a wooden block, a small metal car and a stainless, steel spoon and a CD.

She asked the children to predict which types of matter will be attracted to the magnets.

"I think the magnet will only be attracted to the car and paper clips because they all contain metal," hypothesized Allison.

Everyone agreed except Eli who thought the magnet would also attract the CD.

Everyone tested all of the objects, even some that that were not put out by Farrah's mom.

Since it was Farrah's party, she was chosen to represent the group's conclusion.

"Based on our testing, we conclude that metals are attracted to magnets, but not all metals. The spoon was not attracted to the magnet. We also noticed that the magnet did not even have to touch as object to be attracted to it, though the closer it was the greater the attraction"

"Great summary of what you learned today, particularly that the magnet's force can be felt through a distance. The objects don't even have to touch," added Lisa.

"Mom, Eli and I also noticed when we were placing balloons on the wall sometimes, they did not have to touch the wall. It was like it was attracted to the wall," added Farrah. I guess the force of electricity can also be felt through a distance without touching too."

"Absolutely," responded Lisa proud of her daughter's ability to understand her observations.

"Are you ready to do another experiment?" Asked Farrah's mom.

The cheer that erupted from the table at the mention of another experiment made Lisa smile.

After they had all eaten, Farrah asked, "What experiment are we going to do next?"

"We are going to make a magnet, from an iron nail, some wire and a battery. This type of magnet is called an electromagnet

Lisa sat on one of the stools beside the table and told the children a story. "There once was a scientist named Hans Christian Örsted who lived in Denmark. He accidently discovered that when electricity travels through a wire, a magnetic field is produced. The needle of a compass placed near the wire deflected or moved, when electricity traveled through the wire." Many discoveries in science are made serendipitously (By accident), but a scientist must be an excellent observer.

"Can we make one? Can we make one?" cried all the children excitedly.

Lisa went to the closet to gather materials she purchased for the experiment. She gave each of the children a battery, some wire, masking tape and a long iron nail. She told the children to wrap some of the wire around the nail, but to leave the ends exposed so they could be attached to each end of the battery. Some of the children wrapped a lot of wire onto the nail while others wrapped a little.

As they worked Lisa explained to Farrah's friends, "the wire has insulation on it so the ends of the wire touching the battery have to have the insulation removed in order for the battery to conduct electricity." She gave each of the children some sandpaper to rub off the insulation at the ends of the wire. Lisa told the children to attach one end of the wire to the positive end of the battery and the other to the negative end using the masking tape provided.

Hans Christian Öersted and what was learned

"Look, my compass moves!" some of them called out.

"You have now made an electromagnet," Lisa told them. "Though you cannot see it, a magnetic field surrounds your magnet and it is this field that causes the magnetic needle of the compass to move." She then passed out paperclips. "See if these will stick to your electromagnet."

Some of the children were able to pick up almost all of the paperclips with their magnet, while others were only able to pick up a few.

Lisa looked around and found one of Farrah's friends who had picked up the most paperclips and one who had picked up the fewest. She asked them to both hold up their electromagnets.

"Why do you think Elle's electromagnet can pick up more nails than Noe's?" Lisa asked the group.

They looked at the two magnets.

After a moment, Noe said, "My electromagnet does not use as much wire as Elle's. She wrapped a lot more wire on her nail than I did."

Farrah thought about this observation and added, "Perhaps, the more wire around the nail, the stronger the electromagnet is."

This got all of the children experimenting to see who could make the strongest electromagnet. Eli asked his mom for a very long piece of wire. Sure enough, he was able to make the strongest electromagnet.

"Do you have another experiment?" Farrah asked. So far it had been a lot of fun playing with magnets, balloons and electromagnets.

"Well, I could teach you all how to use magnets and batteries to make a motor," Lisa suggested. This resulted in a great cheer from Farrah and her friends.

"What do we need to make a motor?" Farrah asked.

"You pretty much have all you need to make a simple motor in front of you: a battery, copper wire (precut by Lisa), two large metal paperclips, black magic marker,

magnet, sandpaper and masking tape. Lisa told them what to do, repeating herself as questions were asked. "Wrap some wire around the D Battery about 10 times and let about two inches of the wire to extend from each end. We will call these extensions 'arms.' To hold the wire loop together, wrap extended arm around each side of the loop two times, pulling it tightly. Make sure the wire arms on each end are even with one another. If the looped wire is loose, use some of the tape to hold each half of the wire together. Make sure the wire loop is very round and the arms sticking out are straight and directly opposite one another." Lisa paused and watched as the children followed these directions, providing help when needed.

She continued. "Sand the edges of the wires arms on all sides to remove insulation coating on the wire. Unbend one end of each paper clip so it is straight. Bend the other end down so it can hold each side of the copper wire sticking out from the coil. Attach the straight ends the paperclips to the positive and negative ends of the battery with duct tape. Make sure the paper clip bends are the same height on each end of the battery. Use the magic marker to completely mark one side of each copper wire arm, making sure to mark the same side of each wire."

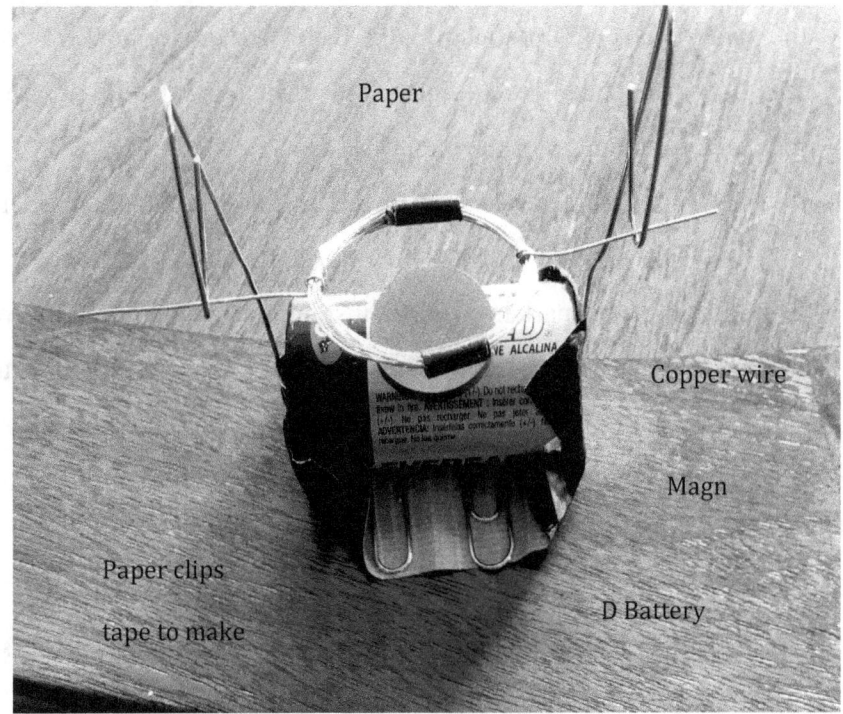

Lisa let the children set up their motors, but intentionally left out some important information to see if they can figure it out.

Farrah set up her motor, but nothing happened. "It doesn't work," she told her mom sadly.

All the children had the same disappointing results. Lisa asked them what could be done to make the motor work based on what they had learned so far.

"Do we need more loops of wire?" asked Noe.

Elle sighed, thinking. "Maybe the batteries are dead."

Farrah shook her head. "I doubt if all the batteries are dead."

"What forces have you learned about to day?" Lisa asked them.

"We learned about electric and magnetic forces," Eli answered.

"Okay, so what happens when electricity travels though a wire?" his mom asked.

"I know," Allison said. "A magnetic field is produced because it can deflect a compass."

"Yes. Now think carefully," Lisa said. "What can you do to push the circle of wire so it turns over the battery?"

It was silent as the children thought.

"Opposite poles of a magnet push against one another," stated Eli.

"Maybe if we place one of our magnets under the wire, it will push against the wire and make it move," Farrah continued.

They tried Farrah's suggestion and most of the children got the wire to turn after they placed the magnet on the top of the battery. The ones that didn't work noticed that the paperclips were not touching the battery or the loops were not round enough. When it was fixed, the coiled wire turns as it is pushed and repelled by the electromagnet and the magnet sitting on top of the battery.

"Would it have worked if we didn't use the magic markers?" Asked Farrah.

"Only until the magnets line up (north attracted to south poles) and stop the wire from turning. The marker provides insulation that temporarily stops the flow of electricity so the coiled wire does not stop. Gravity pulls the magnet down and then the magnetic forces of push or pull kicks in, to keep the coil turning."

Later that evening, after the party was over, each of Farrah's friends went home with a new science toy.

"Thanks mom for organizing all those science activities. My party was so much fun! It was as much fun as the robot we made at Allison's party a few weeks ago!"

DISCUSSION QUESTIONS

1. Explain why and how a balloon can be stuck to a wall.

2. How is the Law of Charges alike and different than the Law of Poles?

3. Explain how to make a motor.

4. Why did some electromagnets pick up more paper clips than others?

5. Why did the life saver magnets move closer together as more magnets were added to the straw?

6. Using these supplies, try out the experiments from this story:

    Balloons
    Lifesaver magnets
    Straws
    Clay
    D batteries
    Copper wire
    Sandpaper
    Paper clips

Farrah's Electric Birthday Party Science Terms

1. **Atoms**: The smallest unit of an element that has the properties of the element.
2. **Conduct:** When an electric current can flow through matter.
3. *Matter*: Anything that has mass and occupies space.
4. **Electric**: Having to do with the movement of electrons.
5. **Electric Circuit**: A pathway through which electricity travels.
6. **Electromagnet:** A type of magnet formed when a current runs through wire wrapped around a magnet.
7. **Electron:** A particle of an atom with a negative charge.
8. **Force:** It is a push or a pull.
9. **Gravity**: A property of matter that attracts other matter. The more matter the greater the pull of gravity.
10. **Insulation**: A type of matter in which current cannot flow.
11. **Law of Charges**: Unlike charges attract while like charges repel.
12. **Magnet**. A type of matter that produces a magnetic field with north and south poles.
13. **Magnetic Field**: The forces that surround a magnet.
14. **Magnetic Forces**: The forces associated with a magnet
15. **Negative Charge**: The net charge on matter when there is more negative than positive charges.
16. **Neutral**: Matter with no charge.
17. **North and South Poles**: The opposite sides of attraction of a magnet.
18. **Nucleus**: The central part of the atom where the neutrons and protons are located.
19. **Positive Charge**: The net charge on matter when there is more positive than negative charges.
20. **Proton**: A particle in the nucleus of an atom with a positive charge.
21. **Repel/Attract**: Forces that push or pull. When something is repelled, it is pushed away. When it is attracted is pulled in.

MS-PS2 Motion and Stability: Forces and Interactions

Disciplinary Core Ideas

S2.B: Types of Interactions

- Electric and magnetic (electromagnetic) forces may be attractive or repulsive, and their sizes depend on the magnitudes of the charges, currents, or magnetic strengths involved and on the distances between the interacting objects. (MS-PS2-3)
- Forces that act at a distance (electric and magnetic) can be explained by fields that extend through space and can be mapped by their effect on a test object (a ball, a charged object, or a magnet, respectively). (MS-PS2-5)

Science and Engineering Practices

1. Asking questions and defining problems
2. Planning and carrying out investigations
3. Constructing explanations and designing solutions
4. **Engaging in agreement from evidence**

Crosscutting Concepts

1. Cause and Effect
2. Stability and change

The Big Debate: Analog (Vinyl) vs. Digital Music

Hunter and Dylan were getting together to listen to some music with their friends, Chessie and Kiley. Dylan was excited about the new vinyl records he purchased.

"Vinyl sound is so much better than the sound from CDs because it uses an analog storage system instead of digital," said Dylan to Hunter.

"The debate between analog vs. digital music has been a round for quite a while," noted Hunter. "Quite honestly, I don't really hear a difference."

"Vinyl is purer because the sound waves recorded are continuous, while in digital music, it is broken up into discrete pieces of data. Therefore, analog sound has to be of a higher quality than digital because it is recording the sound as it is actually happening," replied Dylan. Dylan has a band and played at both high school and middle school functions. His band even recorded a few songs that sometimes play on the local radio.

"I have heard the opposite," said Hunter, challenging Dylan's reply. "Though the sound waves are continuous, it depends on many other factors such as the quality of your stereo equipment. Hunter played guitar in Dylan's band.

Just then the doorbell rang and Chessie and Kiley arrived.

"We are having a bit of a debate over vinyl verses digital music," Dylan said to the girls.

"I prefer vinyl," said Kiley. "The sound is warmer. Those little grooves in the record are laid down exactly as the music is being played. Even when the vinyl gets old and the sound picks up some static, it still sounds better to me.

"Well if someone studied this, I doubt you can measure a factor called, 'warmer.' It is way too subjective," said Hunter.

"Well digital music can accommodate a higher range of sounds, so I think it has the potential to provide higher quality music," added Chessie to the debate. Chessie loved a debate. She was on the 8th grade school debating team.

"When vinyl gets old, the music quality decreases because the grooves in the vinyl degrade. This does not happen in digital music," stated Hunter, feeling pretty sure about his stand.

"Well, to settle this argument, we should do an experiment to see if our friends can distinguish digital from analog music," suggested Dylan.

"We can have a party at my house and have our friends over to test this. I will check with my parents to see if it is possible" said Chessie.

"I have a better idea, said Hunter. "Why don't we test this at school? That way we can get a larger sample of people for testing and the bigger the sample the more accurate our results will be," suggested Hunter.

"Perhaps we can get some extra credit from our science teachers for doing this," added Kiley, smiling.

That is a great idea," replied Chessie. "A little extra credit can always come in handy."

Hunter began thinking about how to go about doing the experiment. "If we are to do a scientific study of this we need to make sure we control certain variables. For example, the CD we use must be recorded from a digital master to ensure it is the highest of quality. The vinyl must be recorded from an analog master.

"Right," said Dylan. "I know that some vinyl records have been recorded from a digital source and CDs from an analog source so we do have to make sure that the masters use the same storage system as what is being tested."

"Hmm, how are we going to get masters of vinyl and digital music," thought Hunter out loud.

"Easy, replied Dylan. "Our music teacher, Mrs. Dippold can help us create masters of both types of recordings.

"That's right, I remember her telling us about it. I am sure she will be happy to help us with this project," said Hunter.

"Now to set up a good experiment, we need to control everything else between the two types of music storage systems just like we do with science experiments," said Kiley, excited about their project.

"Right," said Chessie. "So, we need to make sure that we use the same stereo equipment for both types of music."

"The amplitude or loudness of the sound must be the same," added Hunter.

"How long the music is played must be identical," said Dylan.

"We must play the same musical piece," added Kiley.

"We should probably have a few trials for each participant to help cut back on the chance factor. After all, they still have a 50/50 chance of getting the answer correct," added Hunter.

"Good point," noted Dylan.

"How many participants should we use?" Asked Kiley

"As many as we can get," replied Dylan.

"Well, we don't need the entire school to do this. A sample of the students should work just fine." stated Chessie.

They all agreed that a sample of 25-30 students should be enough.

"OK, now that we agreed on the sample size, we are not exactly experts on sound energy. I think we need to do a little research in this area before we begin," suggested Chessie.

"It is time for a little Googling," said Hunter.

Hunter Googled the following question into the computer: "What is sound?"

Lots of information came up on the computer.

After finding a good, reliable source, Hunter read, "Sound is a form of energy that travels in waves. Unlike light waves (electromagnetic energy), it requires a medium through which to travel. Since the waves cause the medium to compress and then expand as energy is being transported, it is called a compressional wave. Light waves are called transverse waves because the motion of the wave is perpendicular to the direction of the energy being transported. Since sound waves move parallel or in the same plane as the energy being transported, they are also called longitudinal waves."

"Sorta like an accordion," replied Chessie.

"Yes," agreed Hunter. "As a wave is produced, it causes particles in matter to expand and contract. The expansion is called a rarefaction and when the particles move close together, it is called a compression."

"Hey guys, I can feel the sound," said Dylan. "Put your index and middle finger under the front of your neck and say 'aha' softly and loudly.

They all tried it. "That is neat," said Kiley.

"What does a sound wave look like?" asked Chessie.

"Look here said," Hunter who had pulled up a picture of a compressional wave on the Internet.

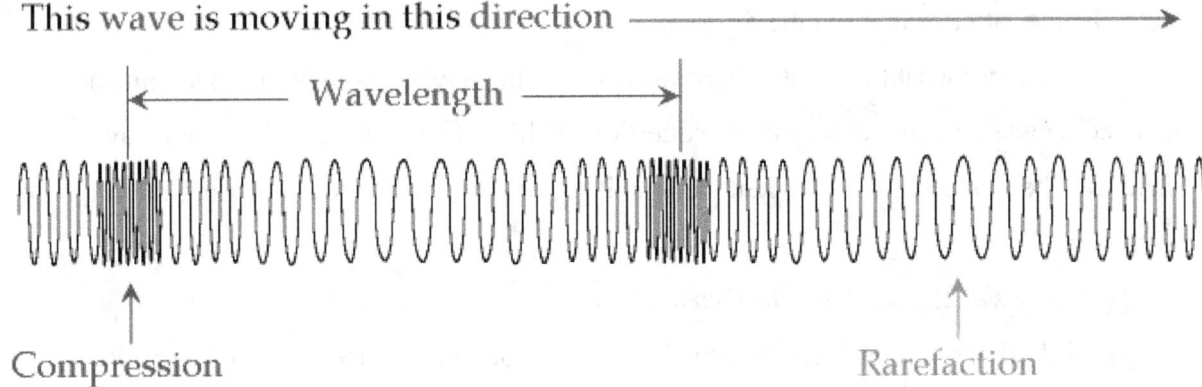

"It says the amplitude of the wave determines how loud the sound is. Looking at the wave, I cannot tell what makes a sound wave loud or soft," said Dylan.

Hunter tried to find a picture that showed the amplitude of a sound wave, but could not find one. Instead he read aloud: "The amplitude of a longitudinal wave is a measurement of the amount of energy being transported. It is a measurement of how compressed or rarified a wave becomes from a resting or undisturbed position (equilibrium). The more energy, the more compressed or rarified the wave becomes and the louder the sound will be. Scientists can measure the size of the amplitude by converting it electronically into a transverse wave.

"Light travels in a transverse wave," stated Chessie.

Hunter showed his friends a picture of a transverse wave with its parts labeled.

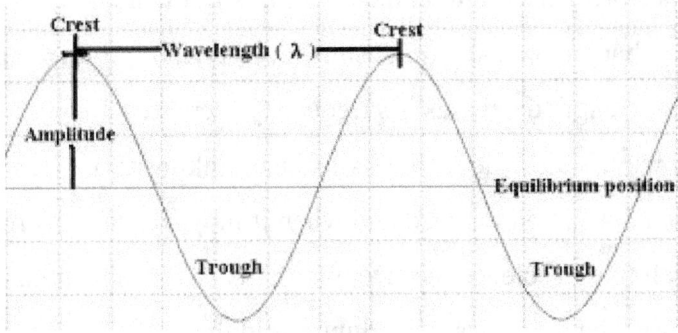

"Ok, I think I now understand the amplitude of sound when it is converted to a transverse wave. It makes a lot more sense to me now," stated Kiley. "

"Yes, it is easier to visualize on a transverse wave. The amplitude is a measurement from crest to a sort of resting point or equilibrium and the higher the amplitude, the louder the sound," stated Hunter after examining the diagram.

"I wonder what the amplitude measures in light," Dylan thought out loud, but soon realized he was able to answer his own question. "Oh! If the amplitude of a sound wave measures the loudness of sound then the amplitude of a light wave must measure the brightness of the light."

"Very swift!" said Kiley to Dylan.

"I think we are getting very knowledgeable about the science of wave energy," complimented Hunter to his friends.

"We still need to discuss the wavelength of a wave," Chessie reminded her friends.

Dylan looked at her and explained, "So looking at the diagram of a longitudinal wave, its wavelength is the distance from rarefaction to rarefaction."

"Or compression to compression," added Chessie.

"We are becoming wave nerds," joked Hunter.

"Aha, so now looking at the transverse wave, I can see the wavelength is from crest to crest or trough to trough," said Kiley, impressed with her own analytical skills.

"Hmm, so what property of a wave causes high and low pitch sounds!" Stated Kiley, feeling quite confident with her ability to understand wave energy.

They all looked at the wave diagrams.

Hunter said, "If I want to hit a high note on my guitar I tighten up on the strings and that causes a decrease in the wavelength and an increase in its frequency."

"We would call that an inverse relationship," said Dylan, proud to apply his mathematical knowledge. "As the wavelength decreases, its frequency increases."

"Don't ask me why," said Kiley, "but I was just thinking about an ambulance. When it comes near you the sound gets high pitched and when it moves away, the reverse happens."

"I think they call that in science the 'Doppler Effect'. As the ambulance moves toward you, it further compresses the sound waves," Hunter told them.

"So, the wave length is shortened," interrupted Kiley.

"Yes," said Dylan, "and when it moves away, the sound waves can spread out more."

"So, the wavelength increases," interrupted Kiley again.

Chessie clarified their understanding. "The 'Doppler Effect,' is just that, an effect. The source of the sound wave has not changed, it is how an observer hears the sound from the motion of the vehicle."

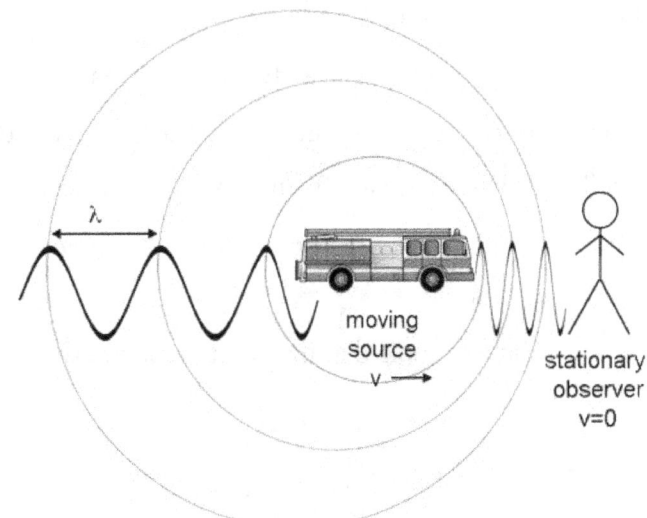

This got Dylan thinking about planes that can go faster than the speed of sound. "The motion of the plane through air creates waves, when the plane catches up with the waves, it is traveling at the speed of sound and the air is compressed. When it goes faster than the speed of sound, the compressed air is released and that is the sonic boom we hear when a plane goes faster than the speed of sound. I am getting brilliant and I am only 14," said Dylan a little smugly, but with a smile on his face.

"And cocky," said Kiley.

"OK, you two, let's get back to our problem. Don't forget that you need to add that frequency is a measurement of the number of vibrations per second and scientists express this in hertz abbreviated Hz" added Hunter. He was on top of his Google search.

"I just looked up the frequency range for the human ear and it is 20-20,000 Hz," said Chessie.

"Right, Google Queen," they all responded.

"Sound technicians must have to know all of this when recording music," said Dylan. Dylan has a friend who is an aspiring singer and works with sound technicians who are friends of his father.

"Did you know that very loud music could puncture an ear drum?" Kiley asked.

Dylan looked at her, confused and amazed. "Now how could that be possible?"

This got them all thinking about what sound waves do to the air before it hits a person's eardrum.

Considering herself the "queen of Google, "Chessie looked up information about this and found that the intensity of sound is measured in decibels, abbreviated db, and loudness is how we interpret the sound. "Decibels can be quantified while loudness is really subjective, depending on how well a person hears." She noted.

"Hey, did you know the decibel is named for Alexander Graham Bell, the inventor of the telephone. Actually, the unit is a bel, but since the unit they use is 10 times a bel, it is called a decibel. It is actually a transmission unit for sound used by the Bell Laboratories beginning in 1928," reported Chessie after reading an explanation from a website on the inventor.

They all looked at the chart on decibels.

Source	Intensity Level
Threshold of Hearing (TOH)	0 dB
Rustling Leaves	10 dB
Whisper	20 dB
Normal Conversation	60 dB
Busy Street Traffic	70 dB
Vacuum Cleaner	80 dB
Large Orchestra	98 dB
Walkman at Maximum Level	100 dB
Front Rows of Rock Concert	110 dB
Threshold of Pain	130 dB
Military Jet Takeoff	140 dB
Instant Perforation of Eardrum	160 dB

(Source http://www.physicsclassroom.com/class/sound/u1l2b.cfm)

After looking at the chart, Dylan said, "It looks like 160 db can damage a person's eardrum and 0 db is when we can first detect a sound.

"Well one of the factors we will not be able to control is that all participants in our experiment on analog (vinyl) vs. digital sound all can hear at the same range," said Chessie.

"That is why we need a good size sample of people doing this," said Hunter.

After working with the music teacher, Mrs. Dippold and creating master music in analog and digital, Chessie, Hunter, Dylan and Kiley organized 30 students to listen to music. They played the same piece of music for the same amount of time 3 times for each subject in their experiment. Each subject listened to the music through earphones and wore a blindfold to eliminate any distractions. After the music was played, a multiple-choice form was provided with 3 choices: DIGITAL   ANALOG   UNSURE.

Hunter offered to compile the results and they set a date to meet and discuss and analyze their data.

The data was organized in a data table as shown on next page.

Subject	Digital	Analog	Unsure	Number of correct answers	Number of incorrect answers
1	√√	X		2	1
2	√	X	0	1	1
3	X	XX		0	3
4	X	√	0	1	1
5	√	√√		3	0
6	X	XX		0	3
7	√√	X		2	1
8	√	√	0	2	0
9	XX	√		1	2
10			000	0	0
11	X√	X		1	2
12	√	√X		2	1
13	√	√	0	2	0
14	XX		0	0	2
15	√	√X		2	1
16	√	X	0	1	1
17	XX	X		0	3
18	√	X	0	1	1
19	√√	√		3	0
20	X	√	0	1	1
21	XX	√		1	2
22	√√	√		3	0
23	XX	X		0	3
24	√	√	0	2	0
25	X	X√		1	2
26	√	XX		1	2
27			000	0	0
28	X	√√		2	1
29	√	X	0	1	1
30	X	√√		2	1
Total			16	38	36
Percent			17.8%	42.2%	40.0%

Key: √ = correct answer   X = incorrect   0 = unsure

A few days later, they all met after school in one of their science teacher's classroom to looked over the results compiled by Hunter.

"Well, of all of our subjects, only two were able to give the correct answer 3 out of 3 times," noted Dylan.

"Sixteen participants or 17.8% of the answers were 'unsure'," said Chessie.

"Well 42.2% did got it correct," said Hunter.

"True, but 40% got it wrong," added Kiley.

"If we flipped a coin, we would have a 50/50 chance of getting heads or tails. It looks to me that the results were like that," noted Dylan

"And there was even a significant number who could not make a choice," said Hunter. If that percentage of students were absent from school due to the flu, the school would say there is an epidemic.

Based on the data they collected, the friends concluded most people could not tell the difference between analog and digital music. However, the fact that some got it correct on all trials made them conclude more studies should be done.

"Perhaps some people can just distinguish sound better than others," stated Kiley.

"That could be true since some people have almost perfect pitch while others are just about tone deaf," added Chessie.

"I think we will let someone else do that study," said Dylan.

They all readily agreed.

A couple of weeks after completing their research on digital vs. analog music, they got together to listen to music at Hunter's house.

"Let's listen to one of your satellite music stations," suggested Dylan.
As they were getting ready to listen to some music, Hunter reminded everyone about how much they all learned by doing the digital vs. analog research.

"It was fun," said Kiley.

"Now we know how we hear the sound coming from the TV and other devices," said Dylan.

"Did you know that Alexander Graham Bell wanted to transmit sound using light energy? In fact, he invented a photophone that used light reflected by mirrors to transmit sound, but it ran into a lot of interference. Since lasers and fiber optics were not invented, he had to settle on electricity for delivery," said Chessie.

"How did you know that?" asked Kiley.

"I read about it when we were doing our research. I thought that was kind of neat, said Chessie. "Remember, I am the 'Queen of Google.'"

"We learned in our research that light also travels in waves," said Dylan. "However, light does not need a medium in which to travel as does sound. That is why we can watch satellite TV."

"Just to make sure you all understand, light is part of the electromagnetic spectrum. We can only see the visible light part of the spectrum, but it also includes waves we cannot see such as infrared and ultraviolet light," said Hunter.

"Infrared is what we feel as heat and ultraviolet gives us a suntan," said Dylan remembering a lesson from his science class.

Hunter showed a picture of the electromagnetic spectrum. They all looked at it, but Kiley was getting a little annoyed.

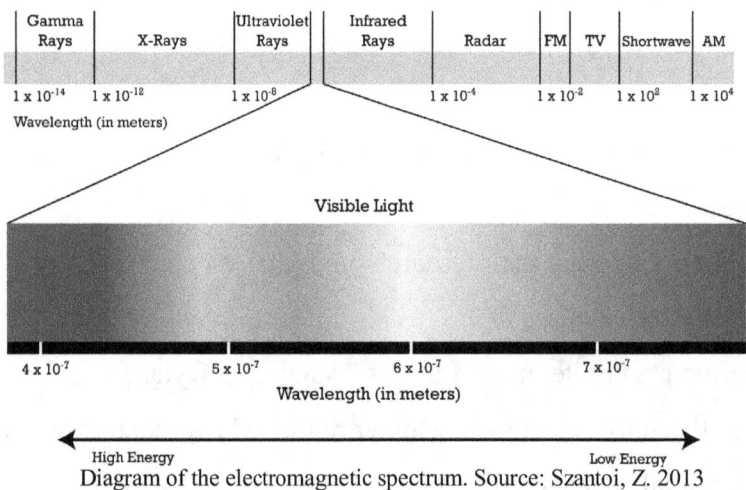

Diagram of the electromagnetic spectrum. Source: Szantoi, Z. 2013

"Light! Sound! I thought we were going to listen to music," said Kiley.

"Hey, I thought you like talking about this stuff," said Hunter.

"I do, but I thought we were done," said Kiley, "OK, if we are going to talk science then who can give me differences and similarities between light and sound waves?" Kiley decided if they discussed this, they could finally listen to some music.

"Do you know the difference?" Asked Dylan.

"As a matter of fact, I do. Our research led me to look up a lot of stuff too," replied Kiley. "Chessie may be the 'Queen of Google,' but my Googling expertise has improved a lot since we began this project. You can call me the "Princess of Google."

But as usual, the 'Google Queen,' Chessie contributed first. "Light travels in a transverse wave while sound travels in longitudinal waves." To remind everyone, Chessie brought up a picture of a longitudinal and transverse wave on the Internet.

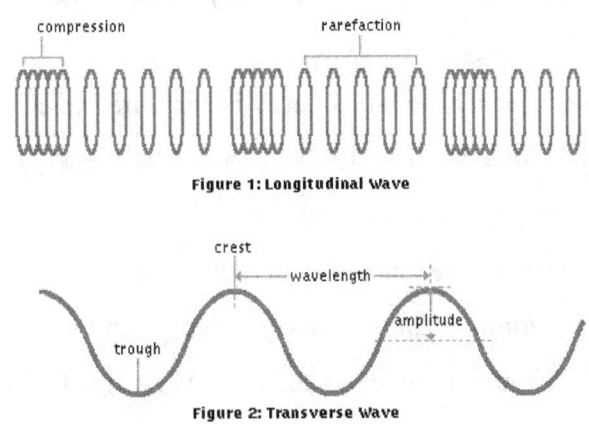

They all looked at the picture. "So, a crest is the same as a compression on a sound wave and a trough is a rarefaction," said Hunter.

"When you increase the amplitude on a light wave, the light gets brighter and, on a sound wave the sound gets louder," said Chessie.

"When we are done analyzing wave energy, we are definitely going to listen to music," Kiley reminded her friends.

"Absolutely," said Hunter, "but don't you want to know how the music gets here."

"Light energy transmitted to a receiver in our stereo system and then it is converted to sound which is transmitted to our ears. So, one property light and sound waves share is that they can be transmitted," said Dylan.

"Another property is that light can be reflected, like when I look in a mirror," said Chessie.

"And sound can also be reflected like in an echo," added Dylan

"OK, we now know that both sound and light can be transmitted and reflected. What else can they both do?" prodded Kiley.

After a couple of minutes of silence, Chessie said, "Light can be absorbed. When I feel hot, I am absorbing the infrared part of the spectrum."

"Sound can too, but some materials absorb sound better than others," noted Dylan. "For example, a sound proof room does a good job in absorbing sound waves."

Kiley decided to summarize for everyone. "So, when we listen to music on Hunter's TV, electromagnetic energy brings the music to the TV. "Kiley stopped and realized she did not know how the electromagnetic energy becomes sound.

Hunter explained. "The electromagnetic energy is converted to electricity that then converts it into sound energy, which we hear. The neat thing about energy is that it can be converted from one type to another," continued Hunter.

Dylan remembered something he learned in his science class. "My science teacher taught us about the *Law of the Conservation of Energy*. It says energy cannot be created or destroyed. So, I guess it just gets changed from one type to another." Dylan was feeling good about how all of this was making sense to him.

"Now that we know sound and light can be transmitted, reflected or absorbed, let's finally listen to some transmitted music," Said Kiley using the scientific term.

Chessie was drinking soda through a straw when she noticed another property of waves. When the straw was leaning to the side of the glass, it looked bent.

"Wait a minute everyone, we forgot another property of waves. They can be bent. Look. My straw looks bent even though it is not. I wonder why this happens?" Asked Chessie.

"Oh no, are we ever going to listen to music!" groaned Kiley.

"Calm down Kiley," said Hunter. "We will listen to music, but since we all did this experiment, some of us are still curious about wave energy."

Chessie did a quick search and announced to her friends, "The term is *refracted* when a wave is bent. And the reason waves bend is because their speed changes when they go through a different medium."

"I know that light waves travel a lot faster than sound waves," In fact, nothing can travel faster than light," said Dylan, proud of something he learned in his science class.

"True," agreed Chessie. She then read from the Internet, "Since sound requires a medium through which to travel, it also depends on the temperature of the medium. For example, in air at 20°C (68°F) sound travels at 1236 km/hr (768 mph). Light does not require a medium, so in a vacuum, it travels about 300,000 mi/sec (186,000 mi/sec).

"You know," said Hunter. "Now that we all have become 'experts' on waves, there is one thing you should realize. Scientists cannot actually see the wave. It is all these properties of waves we have all studied that make scientists conclude sound and light travel in waves."

"We should publish the results of our experiment," said Chessie.

They all agreed to check with the school paper and perhaps put their results on a music blog page.

Finally, they all agreed it was a good idea to listen to the music.

They put on MTVU and a group they never heard of called Imaginary People were playing a song called "Summerstock."

"It makes you want to dance," said Hunter so they all got up and began to dance.

Discussion Questions

1. What is the difference between digital and analog music?
2. Based on their results, were the students able to conclude that analog music was better than digital. Explain your answer.
3. Describe three properties of waves?
4. How is seeing yourself in a mirror a similar property to an echo?
5. Compare and contrast a sound and light wave.
6. Why did the students use a master recording in their experiment?
7. What causes a wave to be refracted?
8. Whey did the students test each person three times?
9. Explain how you can hear and see a television program using the "Law of Conservation of Energy."

Vinyl vs. Digital Science Terms

1. *Amplitude:* Refers to the height of a sound wave or loudness
2. *Analog Master*: Original storage device for analog music.
3. *Analog Storage System*: Information is stored continuously as it is happening.
4. *Bel*: .1 decibels
5. *Compressional Wave*: A form of wave energy that travels parallel to its direction of motion.
6. *Crest:* The highest point of a transverse wave.
7. *Decibels*: Units that measures the loudness of sound.
8. *Digital Master*: Original storage device for digital music.
9. *Digital Storage System*: Information is stored as discrete data.
10. *Doppler Effect*: Waves moving toward an observer, appear to compress while waves moving away expand. For example, the siren of an ambulance gets higher when the ambulance is moving toward you but lower when it is moving away.
11. *Electromagnetic Spectrum* A form of transverse wave energy. It is the type of energy we receive from the Sun.
12. *Equilibrium*: A wave at rest
13. *Fiber Optics*: Pure glass fibers used to transmit light energy.
14. *Frequency:* Measures how often a wave passes through a set point. This is what determined pitch.
15. *Hertz (Hz):* A unit of sound measurement for frequency.
16. *Infrared Light*: A part of the electromagnetic spectrum with longer wave lengths than visible light. Most of the thermal (heat) energy radiated by objects at room temperature are in the infrared part of the spectrum.
17. *Interference*: Variables that prevented the transmission of sound clearly.
18. *Inverse Relationship*: This is when two variables change in opposite ways. For example, the higher the pitch, the shorter the wavelength.
19. *Law of Conservation of Energy*: Energy cannot be created or destroyed only changed
20. *Light Waves*: A wave that travels perpendicular to its direction of motion. It is a transverse wave.

21. *Medium*: The matter through which a wave travels.
22. *Pitch*: Refers to how high or low notes crated by sound.
23. *Quantified*: A variable given numerical values.
24. *Rarefaction*: A part of a compressional wave between compressions.
25. *Refracted*: A property of wave energy that allows waves to bend.
26. *Sample*: Refers to the frequency by which data is stored.
27. *Sound Proof Room*: Room that absorbs sound so it cannot be transmitted out of the room.
28. *Sound Technicians*: Specialists that record sound.
29. *Sound Waves*: A compressional wave that must travel through a medium.
30. *Subjective*: A variable described with words
31. *Transmit*: To send information.
32. *Trough*: The lowest point on a transverse wave.
33. *Ultraviolet (UV)*: A part of the electromagnetic spectrum with wavelengths smaller than visible light. UV light is what gives us sunburn.
34. *Vibrations*: The back and forth motion of matter.
35. *Vinyl*: Uses an analog storage system.
36. *Wavelength*: The distance from one compete wave to the next.

Next Generation of Science Standards (NGSS)

MS-PS4   Waves and Their Applications in Technologies for Information Transfer

Disciplinary Core Ideas

Wave Properties

1. A simple wave has a repeating pattern with a specific wavelength, frequency, and amplitude.
2. A sound wave needs a medium through which it is transmitted.

Electromagnetic Radiation

1. When light shines on an object, it is reflected, absorbed, or transmitted through the object, depending on the object's material and the frequency (color) of the light.
2. The path that light travels can be traced as straight lines, except at surfaces between different transparent materials (e.g., air and water, air and glass) where the light path bends.
3. A wave model of light is useful for explaining brightness, color, and the frequency-dependent bending of light at a surface between media.
4. However, because light can travel through space, it cannot be a matter wave, like sound or water waves.

Information Technologies and Instrumentation

1. Digitized signals (sent as wave pulses) are a more reliable way to encode and transmit information.

Science and Engineering Practices

- Develop and use models
- Develop and use a model to describe a phenomenon
- Obtaining, evaluating, and communicating information

Crosscutting Concepts

- Patterns
- Structure and function
- Influence of science, engineering and, technology on society

It's Elementary

Allison was very excited about her chemistry set that she got for her birthday.

"Let's explode something," said her friend Jordyn as they unpacked the chemistry set.

"Everyone always wants to explode things, I want to test out the chemicals and learn about their properties. If I am going to be a chemist, I need to start early, said Allison.

"I bet you change your mind when you get to college,' said Jordyn.

"I doubt it," replied Allison.

"Ok, let's make sure I got everything that comes with the set."

Allison took out the list of items that come with the chemistry set and began to check off each item, as it was unpacked.

- 50 mL graduated cylinder
- 2, 50 ml beakers and 2, 100 mL beakers
- 2 Celsius thermometers
- 2 pipettes
  2 stirring glass rods
- Metric scale
- Test tube rack and 5, 25 mL test tubes
- Indicator chemicals: iodine, bromothymol blue, pH paper
- Chemicals included: ammonium nitrate (NH_4NO_3), Calcium Chloride ($CaCl_2$), soluble starch $\{(C_6H_{10}O_5)n\}$, glucose ($C_6H_{12}O_6$).
- Safety glasses

"It looks like I have everything," said Allison after they had unpacked.

"What experiment do you want to do first?" Asked Jordyn.

"Well before we do any experiments we should read the information provided with the chemistry set," said Allison.

"When you work with chemicals, there are always safety rules you need to follow, plus it may provide some helpful information," continued Allison.

"Got it," replied Jordyn.

Allison took out the brochure and began to read.

"Always wear your safety glasses when working with chemicals. Though most of the chemicals in this set are not harmful, they can be irritating if they got in your eyes. All chemists always wear safety glasses when they do experiments."

"OK, let's put on our safety glasses," said Allison.

"But there is only one pair that comes with the set," said Jordyn.

"Don't worry, Jordyn," said Allison, "here is one from another chemistry set I used to have."

Allison continued to read from the brochure.

"The alphabet of matter can be found on the periodic table of elements. It is a chart that lists all of the natural and synthetic elements that make up matter. These elements can be found in Earth's lithosphere (rock), hydrosphere (water) and atmosphere (air). They can be found in space, on stars, planets and moons, comets, asteroids. All living things are made up of elements."

"You mean we are all made of chemicals?" asked Jordyn.

"Yup," replied Allison.

"Hmmm, funny, usually when you hear the word chemicals, you think of something that is not good for you, but obviously that cannot be correct because if living things are composed of chemicals all chemicals cannot be bad," deduced Jordyn.

"Right," said Allison. "In fact, some of the chemicals you are composed of came from exploding stars millions and millions of years ago. That is what my science teacher told us"

"Wow," that is neat that we all have a little 'stardust' inside of us," said Jordyn, giggling.

"Does the brochure say anything about the chemicals that come with the set?" Asked Jordyn.

Well, I already know from my science class the chemicals included are all forms of pure matter," said Allison.

Jordyn. "Which do you think is an example of pure matter, a garnet or a rock?"

"Are you giving me a test?" Asked Jordyn.

"Sorta," said Allison, "but I will give you a hint. Pure matter is composed of only one type of matter."

Jordyn took a moment to think about the question. She knew that when she looked at a rock, she saw many different things. It did not look like it was composed of only one thing. A lot of rocks she had collected had some mica in it and other stuff such as quartz. She is familiar with garnets since she has some jewelry made of garnets and it looks pretty much the same throughout.

"I think that garnets are a pure substance and rocks are not because rocks are made up of more than one type of matter," answered Jordyn.

Yup," said Allison, smiling at her friend. A garnet is a mineral and minerals are considered pure substances.

"You sure know a lot of chemistry."

"Well, my parents have gotten me a lot of chemistry sets over the years," said Allison.

"A diamond is a pure substance because it is made of carbon. I remember a Superman movie in which Superman took a piece of coal and crushed it into a diamond for Lois Lane."

"True Jordyn, but most minerals are made up of more than one element. After all, water is not an element because it is composed of oxygen and hydrogen. Good old H_2O!"

"Water is a mineral?

"Yup." Replied Allison.

"Hmm, I wonder what the formula is for a garnet. Allison, let's check the Internet and see what we can find."

After doing a Google search, the girls found the formula for one type of garnet used in jewelry. It is $Fe_3Al_2(SiO_4)_3$"

"What do all of those letters and subscripts mean?" Asked Jordyn.

"Easy," said Allison, "the letters are the chemical symbol for certain elements and the numbers show the ratio in which they combine together. Iron (Fe), aluminum (Al), silicon (Si) and oxygen (O) all chemically combine in a certain ratio (the numbers). Though all of those elements have different properties from one another, when they combine chemically they form a new substance with new properties. In this case they form a type of garnet called almandine according to our Internet search."

"So, elements and compounds are examples of pure substances. A garnet is a pure substance like all minerals. Since rocks are composed of more than one mineral, they are not considered pure substances, but are mixtures," summarized Jordyn.

"I think you got it," said Allison proud that her friend was enjoying chemistry as much as she does.

Allison decided it was time to try some of the experiments in the chemistry kit.

Garnet and rock with garnets in it

"Let's do some solution chemistry," said Allison.

"Time to put on our safety glasses," said Jordyn.

"Now we definitely look like chemists," said Allison, smiling at her friend.

"Or geeks!"

Allison read aloud the introduction to solution chemistry.

"A *solution* is when one substance, called a *solute* is dissolved in another substance, called the *solvent*. A solution can be a solid dissolved in a liquid, a liquid dissolved in a liquid, a gas dissolved in a liquid and a solid dissolved in a solid."

"A solid can be dissolved in a solid?" asked Jordyn with a puzzled look on her face.

"An example of a solid dissolved in a solid is called an alloy. Tennis racquets, baseball bats and jewelry are made from alloys. Alloys provide certain properties such as making a tennis racquet light weight but also strong."

"Now, I get it. That is why gold jewelry is made of 14 carat gold instead of 24 carat gold. Gold is a soft metal and if the jewelry was all gold, it would be too soft, bending and breaking," said Jordyn.

Allison continued to read the directions from her chemistry set.

"Did you ever need a hot pack or a cold pack? When chemicals dissolve they absorb or release heat energy. How do you think a cold pack works?"

Before she read more, Allison thought about the question. "I think I understand. if something gets hot, heat must be released into the environment or how would you know if it was hot! If something gets cold, heat is removed from the environment or why else would it feel cold!"

"I am not sure that I get it," said Jordyn

So, Allison thought about the experiments she had done in her science class and what happens when stuff dissolves. "A cold pack gets cold because during dissolving, thermal energy is taken in from the environment."

"What is thermal energy?"

Allison replied. "It is just a fancy way of saying heat energy."

"Oh, I see, if you remove heat from your surroundings, it gets cold. If you open a door on a cold day, the house gets cold because heat leaves the house. So, when thermal energy is taken in for dissolving, it is like removing heat from the environment. The container in which the dissolving occurs feels cold. Aha! So that is the science behind a cold pack!" Stated Jordyn, cheerily.

"Brilliant! How do you think a hot pack works?" Asked Allison of her friend.

Jordyn felt she was really getting the hang of it so she said, "It must be when a substance dissolves in a hot pack, heat is released and transferred into the environment. That is why a hot pack feels hot."

"Makes sense to me," said Allison, concurring with Jordyn's analysis.

Can the hot and cold packs be reused?" Asked Jordyn.

Not the ones we have here" replied Allison. Once you break the sac inside the pack, they are permanently mixed together. However, I know there are hot and cold packs that are reusable. Those types of cold packs are kept in the freezer. My dad uses one like that when his knees are bothering him. There is also a hot pack that is reusable. My Dad uses that too. His doctor told him to first ice an injury then add heat. So, we always have hot and cold packs in the house. Allison took out a hot pack her father uses and read the label and directions.

"The hot pack comes with a lot of sodium acetate dissolved in it. There is a small metal disk in the sac and if the pack is bent the sodium acetate comes out of solution forming crystals. Since it absorbed heat to dissolve, it releases heat when it crystallizes and so becomes a hot pack. But the neat thing about this type of hot pack is that you can throw it into boiling water and cause the sodium acetate to dissolve again so it can be reused."

"Wow, who would have thought there is that much science in a hot pack," said Jordyn.

Allison read some more from the brochure. "When something dissolves, it is considered to be a physical reaction because no new chemicals are formed. The properties of the substances dissolving are still present. In chemical reactions, new forms of matter appear with new properties.

"Well, it is time to test out our chemicals and see which one is best to use as a hot pack and which one is best to use as a cold pack," said Allison.

Allison massed 5g of calcium chloride. Using distilled water (water with no minerals), she measured out 20 mL of water in the graduated cylinder. Before she poured it into the test tube, she massed the test tube and found it to have a mass of 50 grams. She made a data table to record all of her observations.

"Do you know the mass of the water?" Asked Jordyn.

Allison thought a minute and then made this reply. "Nope, I have no clue. I guess it is time for another Google."

Allison, Googled in mass of water, she got an answer to the "density of water." She learned the density of water is 1g/mL.

She thought it was strange that the density of water popped up, but then she understood.

"Well you do not have to be a rocket scientist to figure out the mass of 20mL of water," said Allison. Since the density of water is 1g/mL, every mL of water has a mass of 1g. Therefore, 20 mL of water has a mass of 20 g."

"Makes sense to me," replied Jordyn.

"What is the density of calcium chloride?" Asked Jordyn.

"I have no idea, said Allison, "but we can look it up."

Allison looked it up on the Internet and found pure calcium chloride had a density of 2.15g/cm^3.

"Why is cm^3 used instead of mL?" Asked Jordyn.

Allison explained, remembering what her science teacher had said. "Usually, in chemistry, scientists use mL to measure the volume of a liquid while they use cm^3 to measure the volume of a solid. However, 1mL has the same volume as 1 cm^3. They are equal (1mL = 1 cm^3)."

Allison poured the 20 mL of water into the test tube. She next massed the test tube with the water and recorded it down as 70g.

Jordyn poured the 5 grams of calcium chloride into the test tube and placed a thermometer into the test tube. Allison looked at the temperature of the thermometer and found it to be 20°C. Jordyn used the glass rod that came with the chemistry set to stir and dissolve the calcium chloride while Allison recorded temperature changes every 5 minutes. She completed a data table.

Time (sec)	Temperature °C
0	20°C
5	25
10	30
15	33
20	35
25	39
30	40
35	40
40	40

Jordyn and Allison then used excel on the computer to graph their data

"Hmm, as I look at the graph it can easily be seen that the temperature goes up and then levels off," said Allison.

"Well if the temperature goes up, then heat must be released," observed Jordyn. Dissolving of Calcium Chloride

"Aha! So that is why the test tube feels hot!" Noted Jordyn.

"I wonder if the mass of the dissolved calcium chloride in water is equal to the mass of the water and calcium chloride before they were mixed together," said Allison.

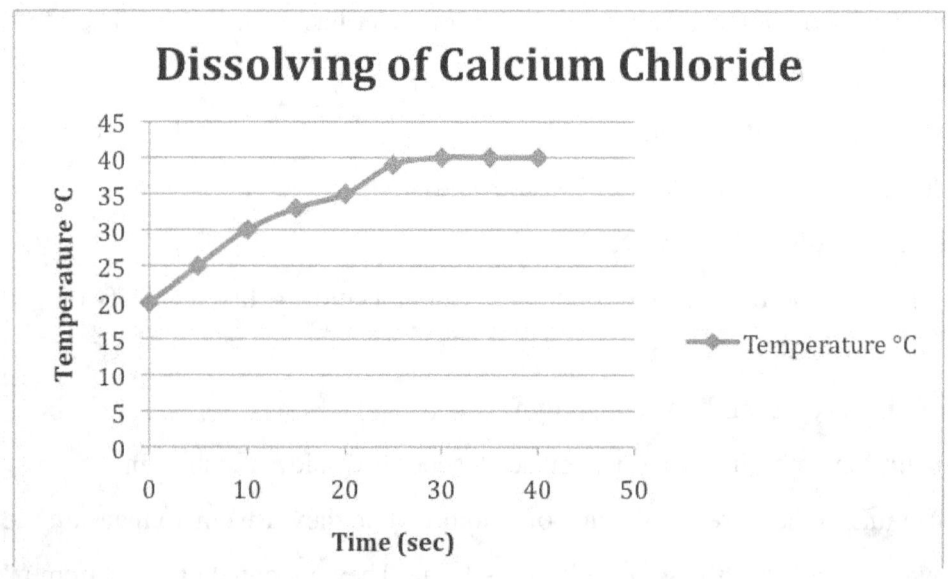

"Let's not forget the important 'Law of Conservation of Matter.' Mr. Hull stressed that in science class. It states that matter cannot be destroyed or created, but it can be transformed into a different type of matter," said Jordyn.

"Just for fun, let's check the law and mass the dissolved calcium chloride and see if its mass equals the mass of the calcium chloride plus water." suggested Allison.

Allison and Jordyn put back on their safety glasses and massed the dissolved calcium chloride.

Jordyn read the reading on the scale to be 75g.

"Don't forget we have to subtract the mass of the test tube," said Allison.

"Then there is 25g of dissolved calcium chloride," said Jordyn.

"And that is equal to the combined mass of the calcium chloride and water before the dissolving,' stated Allison.

"I guess we don't have to worry about breaking any laws today," she joked.

"Are you ready to see what happens to the temperature after we dissolve ammonium nitrate in water," said Jordyn.

Again, Allison and Jordyn carefully placed on their safety glasses. They measured out the same amount of ammonium chloride as they did for the calcium chloride, 5g and the same amount of distilled water was placed in a clean test tube, 20mL.

Jordyn read the starting temperature (before the calcium chloride is placed into the test tube) to be 20°C.

Jordyn placed the thermometer back into the test tube after adding the ammonium nitrate. It was 20°C. She began stirring to dissolve the salt. Allison read and recorded the change of temperature into their data table.

Time (sec)	Temperature °C
0	20°C
5	19
10	17
15	16
20	15
25	12
30	10
35	10
40	10

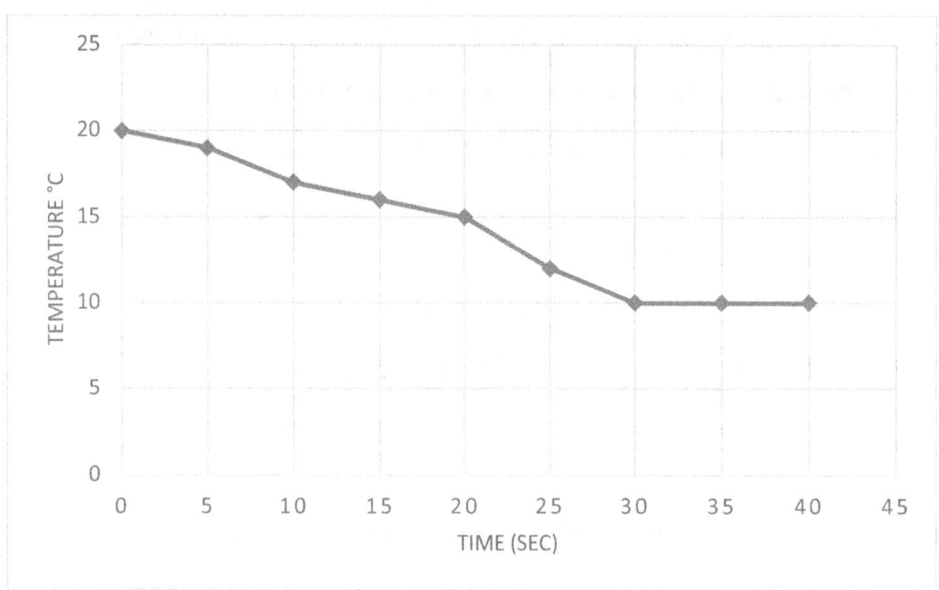

Jordyn and Allison observed that the temperature had gone down and then leveled off at 10°C.

"Well, since the temperature went down, heat energy must have been absorbed when the ammonium sulphate dissolved making the test tube feel cold like in a cold pack, " concluded Allison

"Ok, let me get this clear," said Jordyn looking puzzled again. "When heat is released, the test tube gets warmer, and when heat energy is absorbed, the test tube gets cooler."

"I think you are on your way to becoming a chemist!" Said her friend.

"We know now that matter can absorb or release heat energy, but what exactly is heat energy anyway? Inquired Jordyn.

"Let me Google it." Said Allison.

"According to this website from the University of Vermont, it says heat energy is the energy that is transferred from areas of higher temperature to areas of lower temperature.

"So thermometers measure heat energy,"said Jordyn.

"No,no,no," replied Allison animated now. "My teacher would always get upset about that . Thermometers meaure the energy inside of matter. They measure the average motion or kinetic energy of the particles in the matter. Heat is energy that gets transferred. It is a process that has to take place in order for there to be heat. Scientists say there is no heat inside of matter. It only becomes heat energy when it is being transferred. However, scientists do agree the

that has to take place in order for there to be heat. Scientists say there is no heat inside of matter. It only becomes heat energy when it is being transferred. However, scientists do agree the amount of heat transferred is related to the kinetic energy of the particles in matter and also to how much matter there is. Anyway, that is how scientists define heat or thermal energy. But, I do see how it can confuse some people because if you touch something hot, you think it has heat in it. But, actually, it feels hot because heat is being transferred out of the matter. "

"Believe it or not, I do get it," said Jordyn impressed with ther friend's knowledge. "I think someday you might just become a scientist!" She continued.

"A chemist scientist," Allsion corrected Jordyn.

"This chemistry set is a lot of fun," said Jordyn to Allison. "What do you want to do next?"

Allison decided to learn something with soluble starch. She placed 5g of the starch into a zip loc bag and added 50mL of water. She closed the bag and shook it to it dissolved.

"Let's see if the starch is still there after it dissolves. Physical reactions mean no new substance has been formed. We can test the mixture with iodine. If it turns blue, the starch must be present. Iodine is considered an indicator for starch," said Allison.

"How did you kow that?"

"Mr. Hull's science class!"

Allison was hungry so she suggested to Jordyn they take a snack break. She placed the starch bag in the same beaker she had left the iodine pipette.

When they returned to continue their experiment with the starch, a surprise greeted them.

"Look, look, " exclaimed Jordyn to Allison, the bag of starch turned blue and we didn't even add iodine to the bag!"

"Somehow, the iodine got into the bag. Well, we now know there is still starch in the bag so no chemical reaction took placed when it dissolved so when something dissolves, it is a physical reaction," concluded Jordyn.

Jordyn was curious as to how the iodine got into the bag. "But how did the iodine get into the bag?

Allison looked at the beaker. She knew that iodine turns blue in the presence of a starch. She noticed that the beaker was not completely dry when she placed the zip loc bag in it.

"There must be a leak in the bag," said Jordyn.

However, when Allison lifted up the bag and wiped it off, she could not find any leak.

"Wait a minute," thought Allison, "if there was a leak in the bag, the beaker should have turned blue too, when starch leaked into the beaker, but it is only blue inside of the bag."

"Does that mean the iodine got into the bag?" Wondered Jordyn.

"Well, there is only one way to find out, said Allison. "We will have to do another experiment."

The girls prepared another zip loc bag and added 50mL of water and 5g of soluble starch. They sealed the bag and shook it until all of the starch was dissolved. They then placed it into a beaker that had 25mL of water. Next , using the pipette, they added 3 drops of iodine to the water in the beaker. They made a second set up , but did not add iodine. This was there control so they had a comparison. They carefully observed the bags and sure enough the bag in the beaker with iodine turned purple, while the bag in the water without iodine remained colorless.

Again, they tested the zip loc for leaks and they found none.

"Somehow, the iodine particles must be able to go through the plastic in the zip loc," deduced Allison.

"There must be tiny pores in the bag that allows this to happen, said Jordyn."

"And they must be really, really tiny," said Allison.

Allison recalled what she learned in her science class. She told Jordyn soluble starch mixes into the water breaking up into very small particles, but were too big to go through tiny openings in the zip loc bag.

"The iodine dissolves into iodine molecules which can pass into the bag," concluded Jordyn softly.

They both had learned about molecules and atoms in school. They knew that an atom was the smallest unit of an element and a molecule was the smallest unit of a pure substance. Starch and iodine are both compounds because they are composed of more than one element. Doing the experiment gave them an opportunity to apply what they had learned.

"This had been fun, Allison. "What else do you want to do?

"My teacher made Öoblek in school and it is weird stuff!" Said Jordyn.

"Ok, let's make Öoblek," said Allison.

She looked into the kitchen pantry and found some cornstarch. She emptied the entire box of cornstarch into a bowl and slowly added water as Jordyn mixed it together.

"It is really hard to mix up this stuff," said Jordyn.

"Let's add some green food coloring for fun," said Allison.

Soon the Öoblek was just the right texture. When the spoon was removed, some of the Öoblek oozed off of it slowly. Since Jordyn and Allison had both observed some of the properties of Öoblek, they decided to test it on Allsion's little brother Henry.

"Henry, come on down to the kitchen," Allison yelled.

She told Henry to punch the Öoblek as hard as he can.

"Are you crazy," Henry said. "Why would I do that and have a mess all over me?"

"Ok, said Allison, don't do it.. Jordyn, would you please show Henry what happens when you punch Öoblek real hard."

Jordyn went over to the bowl of Öoblek. She lifted her arm real high and then smacked into the Öoblek. Henry jumped away, but was surprised when nothing splashed out of the bowl.

"How did you do that? Asked Henry.

"It's science," said Allison.

Allison told Henry that you can learn a lot about a type of matter by the way it behaves. You can even "see" what it looks like at the smallest level by the properties it exhibits.

Her teacher did this in science class and Allison had volunteered to punch the Öoblek. He said they should think of Öoblek as looking like strands of sphagetti. The strands get tangled together when the Öoblek is smacked real hard. If you touch it gently, they do not get so tangled, so display the oozing property. She demonstratd this to Henry by softly placing the spoon into the Öoblek and then let the Öoblek ooze off of the spoon.

Jordyn added, "The teacher also said that all matter has positive and negative charges. When the starch is smacked real hard, the long molecules also known as polymers, bend and some of the opposite charges attract one another, holding the starch in place. This too could play a role as to why the Öoblek does not splash out of the bowl when hit."

Henry was more interested in the magic of the Öoblek than the explanations, but he listened carefully and in spite of himself realized he understood more than he thought he would.

Henry decided that he had enough science and went outside to ride his bike. He thought his sister and friend looked pretty funny in their goggles.

"What do you want to do now?" Asked Jordyn.

"We still have not used the pH paper that comes with the chemistry," said Allison.

| 0 | 1 | 2 | 3 | 4 | 5 | 6 | 7 | 8 | 9 | 10 | 11 | 12 | 13 | 14 |

Allison read in the directions that some substances are acids, some bases and some are neutral. The pH scale has a range of 0-14. It looks like this. When dipped into a solution, all they had to do is match up the color change with the scale.

"I know the pH of water is neutral or 7," said Allison. They tested distilled water and sure enough it matched the #7 color on the pH scale.

Allison and Jordyn decided to test a number of things in Allison's home.

"Let's test the water in the creek behind your house," said Jordyn. So they added that to their investigation. They made a chart to summarize their findings and then graphed it on excel.

Item tested	pH
Lemon juice	2
Apple juice	4
Green Tea	9
Tomato juice	4
Carbonated water	2.5
Cabbage juice	8
Baking soda	9
Ammonia water	11
Milk	6.5
Distilled water	7
Soapy Water	9
Bleach	10
Liquid Drain cleaner	12
Creek	5.5

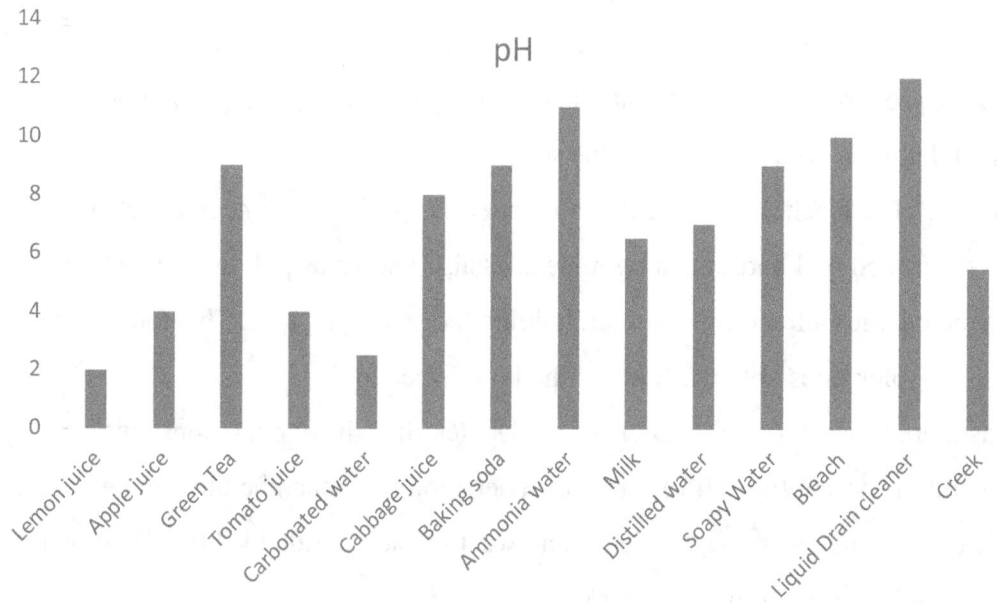

It's Elementary, **Tales of Science** by Joan Wagner

Jordyn said, "It looks like your creek is affected by acid rain."

"Wow," that is very interesting. Allison and Jordyn lived in the Adirondacks of New York State and they knew their area was impacted by acid rain.

"My teacher said that acid rain comes mostly from the burning of coal in the Midwest and the jet stream brings the polluted air to where we live. She said coal is contaminated with sulfur oxides. When the coal burns oxides are released into the air. When it rains or snows, the precipitation is acidic," explained Allison.

"I guess you need to understand chemistry to help the environment too," said Jordyn.

"I know that some of the lakes in the Adirondacks are limed to help neutralize the lakes," said Allison. I will have to ask my parents if we should lime the creek," she continued.

"Why are they limed," asked Jordyn.

"Lime is a base with a pH greater than 7 so it can neutralize the water." Answered Allison.

"We have not used the bromothymol blue yet," said Jordyn.

Allison looked at the directions in the chemistry kit and it said to mix baking soda in some water and test it with bromothymol blue. If the solution turns yellow, it means that carbon dioxide is produced."

Allison looked in the pantry and found some baking soda.

"Put on your safety glasses," said Allison to Jordyn.

They measured out 50mL of distilled water and placed it in one beaker and then set up a second beaker with another 50mL of distilled water.

"Why do we always use distilled water instead of tap water?" asked Jordyn.

"Tap water has minerals in it that can bias our results. We want to make sure that all of our results are due to the chemicals we used," replied Allison.

Using the pipette, Allison added 5 drops of bromothymol blue to both beakers of water.

Next, she added 5g of baking soda to the water in one of the beakers.

"Look at the bubbles, a gas is being produced," observed Allison.

"And the water is turning yellow," added Jordyn.

"So what gas do you think we produced?" Asked Allison.

"Easy," answered Jordyn. "It must be carbon dioxide because bromothymol blue undergoes a color change in the presence of carbon dioxide.

"Give me a high five," said Allison. "Indicators are very useful tools to scientists like us," she continued.

"Do you think this reaction was a physical or chemical reaction? Allison asked Jordyn.

Jordyn thought for a moment and then said. "Since there was no carbon dioxide present before the baking soda was added to the water, it must be a chemical reaction because something new was produced."

"I wonder if heat was released or absorbed," thought Allison out loud.

"Let's do it again and take the temperature this time," suggested Jordyn.

When they repeated the experiment, they learn that heat was released because the temperature did rise, though only slightly.

"What other chemical reactions can we try?" Asked Jordyn.

"We better clean up the mess from both test tubes fizzing out of the test tube first."

After the mess was cleaned up, Allison thought about how her teacher had them make a substance she called Gunk and Glube.

"Let's make Gunk and Glube," said Allison.

Allison looked through her school notebook and found the ingredients needed for both experiments. This is what she found:

Gunk

100mL Elmer's Glue

100g non-iodized salt

100mL Liquid Starch

The Directions said to try and make a type of matter with a high bounce by mixing different amounts of the chemicals. It does not say how much to use

Glube

100mL Elmer's Glue mixed with 100mL water

25g Borax mixed with 100mL water

1quart size zip loc bag

Directions: Pour 150mL of Elmer's glue mixture into a zip loc bag. Next add 50mL of the borax solution to the bag. Seal the bag and mix together.

Allison and Jordyn put back on their safety glasses. First, they did the Gunk experiment. They were very proud of the bouncy balls they made. They wrote down how much of each chemical they used to get the bounciest balls. It is their secret so if anyone wants to make a bouncy ball, they will have to experiment by trial and error.

The Glube was a lot of fun because they created a type of slime. They both loved the way it oozed.

"It must be another polymer," said Jordyn since she observed it to have similar properties to the Öoblek.

Since Glube and Gunk produced substances with new properties, they were examples of chemical reactions.

"We sure have done a lot of experiments today," said Jordyn.

"I am glad you were able to play with my chemistry set with me today."

"We didn't even have to blow up anything to have fun," joked Jordyn.

"I have a quiz question for you," said Allison to Jordyn.

"Name a type of matter that can exists naturally in three phases on our planet."

Jordyn thought for a moment and then replied, "Easy, water exists as ice in places like the Arctic and Antarctica, as a liquid in our lakes and oceans and as a gas in the air.

"How do you know there is water vapor in the air? I do not see any water vapor, said Allison.

"Well, how else would there be rain or clouds?" Replied Jordyn.

Allison decided to Google water vapor.

According to Google, Allison read, "water vapor is a colorless and odorless gas so you cannot see or smell it. Clouds form when water vapor in the air undergoes a phase change to a liquid or solid. Clouds can be made of water droplets or ice crystals, depending on the temperature."

"I remember doing phase change experiments in Mr. Hull's class, noted Jordyn.

"He said that all matter can undergo a phase change with the right temperature. Mr. Hull said it is so hot on Venus that lead can melt on it!"

"Now that is hot!" Said Allison.

"Do you know what it is called when it changes from a solid to a gas," asked Allison of Jordyn.

"Got me on that one," she replied.

"It is called *sublimation*," answered Allison proud that she is remembering a lot from her science class.

"Sublimation, sublimating, sublime." Fun word said Jordyn.

"What is the phase change when ice melts?" Asked Allison

"Duh, a trick question because the answer is *melting*," replied Jordyn.

"What is the phase change when liquids change into a gas?" Asked Jordyn.

"*Evaporation*." Jordyn and Allison said together.

"Now for the tricky one, what is the phase change called when a liquid becomes a solid?"

"It's *freezing*," they both replied together laughing.

"We're on a chemistry roll!"

Allison asked, "What type of phase change occurs when water vapor becomes a liquid like in clouds?"

Neither of the girls responded.

"I thought you knew," said Allison.

"I thought you would know," replied Jordyn. "After all, it is your chemistry set."

"Time for a Google."

Allison learned that when water vapor becomes a liquid the phase change is called *condensation*. She also learned that all matter can change phase with the right temperature.

"Wait! Wait!" called out Jordyn to Allison. "Does it say anything about a phase change when a gas changes into a solid? You know, the opposite of a solid to gas."

"Let me see," replied Allison. Allison read some more from her Google search and learned that when a gas changes directly into a solid, it is called *deposition*.

"So, water vapor can change directly into snow. It must be really cold for that to happen." Concluded Jordyn.

"Same thing must happen when frost forms on the outside of a glass of ice water," said Allison agreeing with Jordyn's conclusion.

"Obviously a phase change is a physical reaction because no new matter is formed."

"Right," agreed Jordyn. "Water, ice and water vapor are all the same matter with the same properties.

"I think it is time to clean up," said Allison. "My parents don't want to see a messy 'lab'."

Allison and Jordyn cleaned up the kitchen and all of the equipment they used.

"We will have to try some new experiments another day Allison said to Jordyn."

"Perhaps we can make just a little explosion next time, Jordyn joked. "Just a real little one!"

Discussion Questions

1. What safety precautions did Allison and Jordyn take when doing their experiments?
2. Why did the girls think that a diamond is a pure substance?
3. Explain how a cold pack works.
4. Explain how a hot pack works.
5. Why did the girls conclude that a phase change is a physical reaction?
6. The girls did an experiment using iodine and liquid starch. Why did the starch turn color?
7. In the experiment with starch and iodine, the girls had two set-ups. What was the purpose of the set-up that did not have iodine?
8. Is dissolving a physical or chemical change? Explain.
9. The girls made Glube, Gunk and Öoblek. Though Glube and Gunk are chemical reactions, Öoblek is not. Explain.

IT'S ELEMENTARY SCIENCE TERMS
1. **Acid**: A substance with a pH of less than 7.
2. **Acid Rain**: Rain with a pH of less than 7.
3. **Atmosphere:** The layer of gases surrounding Earth.
4. **Base**: A substance with a pH of more than 7.
5. **Chemical Reaction**: A reaction in which the substances lose their properties to form a new substance with new properties.
6. **Compound:** A pure substance composed of one or more elements chemically combined.
7. **Condensation:** A phase change from a gas to a liquid.
8. **Control**: It is the part of an experiment that does not received the factor being tested.
9. **Crystallizes**: To become a solid that is made up of regularly repeating particles.
10. **Density**: It is the amount of mass per unit of volume.
11. **Deposition**: A phase change from a gas to a solid.
12. **Dissolves**: When one substance becomes evenly distributed throughout a second substance. The substance being dissolved is called the solute while the substance doing the dissolving is caused the solvent.
13. **Distilled Water**: water that has all dissolved minerals removed from it.
14. **Element:** A pure substance composed of only one type of atom.
15. **Energy:** The ability to do work.
16. **Evaporation:** A phase change from a liquid to a gas.
17. **Freezing:** A phase change from a liquid to a solid.
18. **Garnet:** a group of silicate minerals used as a gemstone and abrasive.
19. **Heat:** Thermal energy. The type of energy that moves from an area of greater temperature to one of lesser temperature.
20. **Hydrosphere:** The part of Earth covered with water.
21. **Jet Stream**: A major current of air moving from west to east around Earth.
22. **Kinetic Energy**: Energy of motion.
23. **Law of Conservation of Matter**: The scientific principle that matter cannot be created or destroyed. In any chemical reaction the mass of the reactants (substances reacting) must be equal to the product (substance formed).
24. **Limed**: a process to neutralize acid water with a basic substance.

25. **Lithosphere:** The solid, outer rocky layer of Earth.
26. **Matter**: Anything that occupies space and has mass.
27. **Melting**: A phase change from a solid to a liquid.
28. **Mica:** A mineral common is many types of rocks.
29. **Mineral:** A naturally occurring, pure substance.
30. **Mixtures:** Two or more pure substances not chemically combined.
31. **Molecule**: The smallest unit of a pure substance composed of two or more atoms.
32. **Negative Charge**: When matter has more negative Charges than positive charges.
33. **Oxides**: A substance that is chemically combined with oxygen.
34. **Periodic Table of Elements**: List of elements developed by Mendeleev.
35. **Phase Change**: A physical change when matter changes its state.
36. **Physical Reaction**: A reaction in which the substances maintain their individual properties.
37. **pH Scale**: Measures the strength of an acid or base solution.
38. **Polymer**: A substance composed of long-chained molecules.
39. **Positive Charge**: When matter has more positive charges than negative charges.
40. **Pure Substance**: A type of matter composed of only one type of substance.
41. **Precipitation:** When a substance settles out of a solution. For example, when it rains, water droplets in the atmosphere settles out.
42. **Quartz:** It is the second most common mineral found on Earth after feldspar.
43. **Solute:** The substance being dissolved in a solution
44. **Solution**: A homogeneous mixture in which one substance is dissolved in another.
45. **Solvent**: The substance in which a solute, dissolves.
46. **Solution Chemistry**: The study of what happens when one substance dissolves into another.
47. **Sublimation**: To change from a solid to a gas.
48. **Thermal Energy:** The type of energy that moves from an area of greater temperature to one of lesser temperature.
49. **Water Vapor**: the gas phase of water.

NGSS Standards

Structure and Properties of Matter
1. Substances are made from different types of atoms, which combine with one another in various ways. Atoms form molecules that range in size from two to thousands of atoms.
2. Each pure substance has characteristic physical and chemical properties (for any bulk quantity under given conditions) that can be used to identify it.
3. Gases and liquids are made of molecules or inert atoms that are moving about relative to each other.
4. In a liquid, the molecules are constantly in contact with others; in a gas, they are widely spaced except when they happen to collide. In a solid, atoms are closely spaced and may vibrate in position but do not change relative locations.
5. Solids may be formed from molecules, or they may be extended structures with repeating subunits (e.g., crystals).
6. The changes of state that occur with variations in temperature or pressure can be described and predicted using these models of matter.

Chemical Reactions
1. Substances react chemically in characteristic ways. In a chemical process, the atoms that make up the original substances are regrouped into different molecules, and these new substances have different properties from those of the reactants.
2. The total number of each type of atom is conserved, and thus he mass does not change.
3. Some chemical reactions release energy, others store energy.

Definitions of Energy
1. The term "heat" as used in everyday language refers both to thermal energy (the motion of atoms or molecules within a substance) and the transfer of that thermal energy from one object to another. In science, heat is used only for this second meaning; it refers to the energy transferred due to the temperature difference between two objects.
2. The temperature of a system is proportional to the average internal kinetic energy and potential energy per atom or molecule (whichever is the appropriate building block for the system's material). The details of that relationship depend on the type of atom or molecule and the interactions among the atoms in the material. Temperature is not a direct measure of a system's total thermal energy. The total thermal energy (sometimes called the total internal energy) of a system depends jointly on the temperature, the total number of atoms in the system, and the state of the material.

Developing Possible Solutions

1. A solution needs to be tested, and then modified on the basis of the test results, in order to improve it.

Optimizing the Design Solution

1. Although one design may not perform the best across all tests, identifying the characteristics of the design that performed the best in each test can provide useful information for the redesign process—that is, some of the characteristics may be incorporated into the new design.
2. The iterative process of testing the most promising solutions and modifying what is proposed on the basis of the test results leads to greater refinement and ultimately to an optimal solution.

The Case of the Ball that Would Not Bounce

One day, Tiah and Amelia found a strange ball. When they dropped it, they found that it would not bounce like other balls they had encountered. Tiah was ready to place it in the recycle bin but Amelia stopped her.

"Wait! Don't throw it away yet. Let's see if we can figure out why the ball won't bounce first."

"We're not doing an experiment, are we?" Asked Tiah, with a little attitude.

Amelia grinned and nodded.

"But that is what we do in school, Amelia. You have to be kidding," said Tiah desperately.

"No Tiah, trust me, this will be fun," replied Amelia to her best friend.

"Whatever," responded Tiah with her favorite answer.

The two girls walked into Amelia's room and sat on her bed. As they examined the ball, they noticed that it felt and looked like any ordinary rubber ball, so why didn't it bounce?

"OK, Amelia, there is nothing we can tell by how the ball feels so let's dump it and Facetime Tonya," said Tiah starting to get bored.

"Don't you think it is weird that the ball feels exactly like any ordinary rubber ball but does not bounce? Come on Tiah, it is rainy outside anyway, it can be fun to find out why it doesn't bounce," replied Amelia, slightly annoyed her friend does not share her interest.

"Whatever," replied Tiah.

"Let's see what other balls we can find," suggested Amelia.

"Why?" Asked Tiah.

"You'll see," replied Amelia.

So, they went around Amelia's house and found several different balls.

Hey, Amelia, look what I found under your sister's bed, it's one of those high "bouncing balls," said Tiah, as she bounced the ball and watched it almost hit the light in the ceiling.

"Whew, close call. My parents wouldn't have been happy if we broke the light," said Amelia, as she watched the ball almost become a problem.

The girls took inventory of the balls they collected: a large soft rubber ball, medium size rubber ball, high bouncer, tennis ball, ping pong ball, golf ball and a plastic ball

"Wow, you sure have lots of balls in your home," noted Tiah. "Don't tell me you want to drop all of the balls," she continued with a touch of sarcasm in her voice.

"Yup," replied Amelia, "And I will give you the honors of starting the experiment."

"Whatever," replied Tiah and she reluctantly dropped the ball first.

"Notice how the ball bounces lower and lower after each bounce," said Amelia, excitedly.

"What's so strange about that, all balls do that," replied Tiah.

"But why does it bounce lower and lower until it cannot bounce anymore? Asked Amelia.

"Perhaps, it just gets tired," replied Tiah.

"Very funny," said Amelia, rolling her eyes. Now think, Tiah, if we can answer that, we can find out why the ball we found can't bounce," Amelia continued with a determined look on her face.

"Hmm, OK, now you are making me a little curious," said Tiah. "Let's take turns dropping the remaining balls. Since I went first, it is your turn, Amelia,"

See, this is fun," said Amelia feeling very determined to discover why the ball they found would not bounce.

"Whatever," repeated Tiah, though a little more softly.

After all of the balls were dropped, Tiah then said, "well, no surprise, all the balls bounced lower and lower until they no longer could. Just like the first ball, they too got "tired" and stopped bouncing, though none of them bounced as long as the high bouncer. Boy, that ball must have been really tired when it stopped bouncing,"

"Balls can't feel anything," said Amelia, they just have less energy to bounce.

"When I feel tired, I have less energy so it just made me think the balls have less energy too," said Tiah.

"Well you might be on to something there, but we can get back to the energy stuff later" replied Amelia. "For now, let's concentrate on what we can observe about a bouncing ball. What else do you notice when a ball bounces?"

"Beats me," replied Tiah, trying to hold on to three balls.

Just then the ping pong ball fell out of her hands.

"Hey, did you hear that? Said Tiah. "I have an idea, let's listen to the sounds each ball makes when it bounces" she continued, surprised she was actually getting into experimenting.

"Whatever," replied Amelia. "Oh, just kidding, you have a great idea. Let's take turns again dropping each of the balls and listening to the sound each one makes."

The girls decided to create a chart (as they were taught in school) that listed all of the balls, the number of bounces, how long they bounced, in addition to the sound each ball made.

"The balls sure do make different sounds, stated Tiah after listening to all of the balls.

"Yeah," replied Amelia, "Some are high pitched while others are low pitched. The large rubber ball makes a loud, hard sound while the tennis ball makes a softer sound," She continued.

"You know it could be fun to play a game in which you had to guess the type of ball by listening to it bounce," suggested Tiah

"Tiah, interesting, but let's stick with our task, please," Amelia pleaded.

"Okay, Okay and when they invent a famous APP game on bouncing balls, you will be the reason we didn't get famous," replied Tiah, wishing her friend would chill out a bit.

Amelia, determined to keep her friend on task, asked her which balls had the lowest pitched sound.

"You will see I did pay attention, Amelia, "the squishier balls made a lower pitched noise than the others. Honestly, I am interested in solving the mystery of the ball we found that forgot how to bounce. Maybe it's sick and needs to see a ball doctor. Oops, just kidding Amelia. It's just you get so serious about it. I am trying to lighten things up a bit," Tiah continued trying to explain her behavior to her friend.

"No problem, Tiah, "you are still my best friend. Now let's look over our chart and the data we have collected so far for each of the balls," Amelia suggested.

"I agree, it is a good idea to see what info we have so far, replied Tiah.

"You know, Mrs. Tuchman, our science teacher would be very proud of the chart we created," said Amelia, impressed with the data they had collected thus far.

The chart showed some of the balls bounced longer than other and the sound was noted as high, low or medium pitched.

Ball Type	Number of bounces	Time	Sound pitch
High bouncer	10	4.48	low
Large Rubber	3	1.93	Medium
Tennis	6	2.83	medium
ping pong	15	8.10	high
golf	8	3.46	high
Medium rubber	8	3.38	Medium
Plastic	9	4.03	high

"While holding a ping pong ball and rubber ball in her hand, Tiah said, "Let's see if the squishiness of a ball affects its bounce. The rubber ball certainly is squishier than the ping pong ball"

"Brilliant," responded Amelia, pleased her best friend was getting into experimenting.

Tiah sat on the floor with their chart in her hand and added the trait of "squishiness" to it. She was going to observe and record each type of ball when it touched the floor after Amelia dropped them. Tiah noticed that the squishier a ball was, the more it changed shape when it hit the ground and when it bounced, it would change back into its original shape.

"I am beginning to feel like a scientist, "Tiah said as she recorded how the balls changed shape when they dropped. "But this is a hard observation. I don't really notice any change with the ping pong ball, plastic ball and the golf ball," she continued.

"Well just write down what you can see and 'no change' when nothing is noticed," suggested Amelia."

"Check, will do," replied Tiah, trying to sound like a scientist.

After recording all the data about the balls, Tiah stood up so they both can look over their findings on the chart.

Ball Type	Number of bounces	Time	Sound pitch	Squishiness (Shape change)
High bouncer	10	4.48	low	Small
Large Rubber	3	1.93	Medium	Small
Tennis	6	2.83	medium	No change
ping pong	15	8.10	high	No change
Golf	8	3.46	high	No change
Medium Rubber	8	3.38	Medium	Small
Plastic	9	4.03	high	No change

"Well, each ball certainly has different properties," Tiah noted.

"They sure do," replied Amelia with a confused look on her face.

Sensing this, Tiah said, "What's wrong?"

"Well we have collected a lot of information about different balls, but I still don't get why a ball loses its bounce over time and why the ball we found cannot bounce at all," replied Amelia, feeling very frustrated.

This made Tiah think about her trampoline at home. "You know when I jump on my trampoline, I can bounce for a while on it without jumping. But, eventually I stop bouncing unless I jump again."

"Are you sure it is not due to your getting tired like the tired ball," replied Amelia grinning.

"I guess I deserved that reply, after all the kidding you had to take from me," said Tiah.

Amelia thought this over and yelled, "I got it, I got it. Jumping on the trampoline is sort of like bouncing a ball. If I want the ball to bounce again after it stops, I have to drop it down again. If I want to continue bouncing on a trampoline, I have to jump up again and then each time I hit the trampoline, it pushes me up."

"Sooooo," said Tiah puzzled.

"Well if the trampoline pushes me back up then the floor must push the ball back up," said Amelia, getting very excited.

"Really," replied Tiah. "A floor can push?"

"Tiah, place your hand on the floor and push down," directed Amelia. What do you feel?"

Tiah pushed against the floor. "I feel the floor."

"Keep pushing harder and harder," said Amelia.

"Ouch, it is beginning to hurt," said Tiah and she took her hand off the floor.

"Don't you see," said Amelia, "the floor is pushing back at you and the harder you push, the harder it pushed back."

"Definitely weird, but I think I get it," Now I know why a ball bounces higher when it is thrown at the floor instead of being just dropped. The harder the ball hits the floor, the harder the floor pushes back," said Tiah, smiling since she was really getting into solving the mystery of the ball that doesn't bounce. Then Tiah said, "Maybe the reason the ball didn't bounce was because we didn't throw it at the floor."

Tiah threw the ball that would not bounce on the floor but just as before it did not bounce.

"Do it again," said Amelia wanting to make sure they observe the same thing.

Tiah threw the ball that would not bounce three more times on the ground and it still not bounce.

"Well that kills that prediction," Tiah said with disappointment in her voice.

"Scientists would call it a hypothesis," said Amelia trying to educate her friend.

"Oh yeah, I remember that from school," said Tiah.

"Don't feel bad Tiah, lots of hypotheses turn out not to be true. We just need to gather more data about bouncing balls," said Amelia, trying to comfort her friend.

"I know what we can do next," said Tiah brightening up, "we could find out how long a ball bounces when it is dropped from different heights. It might help us find an answer to the problem," continued Tiah.

"Great suggestion! We can drop the balls from different heights using the stairs, but we better watch out for the lamp below," said Amelia.

"No problem," replied Tiah.

"Should we count the number of bounces before the ball stops or how long it takes the ball to stop bouncing?" Asked Tiah.

"Both," answered Amelia. "You count the number of bounces and I will keep track of the time with a stop watch."

Tiah made a new chart showing the height the ball was dropped, the number of bounces and the time it bounced. They repeated this three times for each ball and recorded the average.

Ball Type	Height 1 7.9'	Time 1 (Sec)	Height 2 5.10'	Time 2 (Sec)	Height 3 4.3'	Time 3 (Sec)	Average Time (Sec)
High bouncer	12	6.30	10	5.51	9	5.00	5.57
Large Rubber	4	2.56	3	2.00	2	1.87	2.14
Tennis	7	3.78	6	3.63	5	3.12	3.51
ping pong	18	9.20	15	8.23	14	7.69	8.37
Golf	24	12.75	22	11.41	20	9.65	11.27
Medium Rubber	10	5.35	9	4.45	8	4.16	4.65
Plastic	14	6.05	12	5.96	10	5.21	5.74

The Case of the Ball that Would Not Bounce, **Tales of Science** by Joan S. Wagner

Using the stairs, as they expected, the higher the ball was dropped, the longer it bounced and the more it bounced.

The girls looked over their data that they placed on their chart so they can see how the balls compared.

Tiah spoke first. "You know it is sort of weird to watch a ball bounce. Obviously, gravity is what pulls the ball down, but when it bounces up, it's as if it is defying gravity."

"Yeah, if you think of it that way, it is weird except we already determined that the floor pushes the ball up and the strength of the push is determined by how hard the ball hits the floor. Remember what we did earlier!" Amelia said.

"I do, I do, but it is still weird. Normally, no one ever thinks about that but because you decided to solve the mystery of the ball that will not bounce, my brain is on overdrive," said Tiah, looking a bit frazzled.

"Now to summarize what we have observed. Gravity is the force that pulls the ball down and the floor provides the upward pushing force," concluded Amelia acting like she was a teacher summarizing a science lesson at school.

"And, according to our data, the higher the ball is dropped, the longer it bounces. Then again, I am not exactly surprised by that," noted Tiah.

"But do you know why it bounces longer?" Asked Amelia.

"Because it becomes more energetic the higher up it is dropped," answered Tiah, thinking she was just joking.

"You are right," said Amelia. "I told you we would get back to the energy thing you mentioned earlier."

"I am!" answered Tiah in disbelief. She knew science was not her best subject in school and often depended on Amelia for some help.

Amelia explained. "The higher the ball is dropped, the more energy it can store up. That is why it can bounce longer"

"We did study energy in school. What are those terms we learned in Mrs. Tuchman's class?" Asked Tiah.

Amelia thought a moment. "Mrs. Tuchman said that stored energy is called potential energy, and if it has to do with gravity, such as our bouncing balls, then it is called gravitational potential energy."

"Really, how do you remember that stuff?" Asked Tiah impressed with her friend's ability to recall information like that.

"I don't know, I just do," answered Amelia.

Then Tiah started to smile. "You know, when you think about all this it's as if the ball is running out of the energy it needs in order to bounce."

"Yup, it gets exhausted as you would say and just stops bouncing because it has run out of the energy needed to move," replied Amelia.

"What do you call that motion energy we learned about in school, no, no wait, I think I remember, it is kinetic energy. My mother goes to a kinetic exercise class and obviously, they do a lot of moving, so that is why I can remember that word" said Tiah.

"Let's review what we learned about bouncing balls," stated Amelia, always a bit of the teacher in her.

"Okay, so when the ball bounces up, it is storing energy, also known as gravitational potential energy, not bad, huh, Amelia," said Tiah proudly because she remembered the term.

"I'll make a scientist out of you yet," replied Amelia.

Amelia continued, "So when the ball bounces as high as it can, that must be the maximum amount of stored energy it can store."

"This is actually making sense to me," said Tiah in disbelief. "And as the ball falls down, its potential energy is changed back into energy of motion or kinetic energy. Oh no, I am becoming a nerd like you Amelia," said Tiah, with a broad grin on her face"

"Welcome to my world," replied Amelia. "So now we both know a bouncing ball stores up energy when it moves up and releases it as kinetic energy when it falls down."

"And since it bounces lower and lower, less and less energy is stored so when it can no longer store up energy to bounce, it stops bouncing. Poor ball, must feel sad not being able to bounce anymore," noted Tiah.

Tiah, why are you always giving feelings to balls? They only thing they 'feel' is the push from the floor when it is bouncing," said Amelia, a little annoyed that her friend continues to give feelings to a nonliving thing.

"Amelia, you really need to lighten up. I like to give emotions to the ball, it makes doing this more fun," replied Tiah, a bit annoyed.

"Sorry Tiah, I do have to lighten up," said Amelia to Tiah, "Speaking of fun, remember the good times we had on the slide when we were in elementary school?

"Yeah, it was fun, sliding down together or going face down. Though I do remember going down the slide too fast and ending up in the mud. What made you think about the slide?" Asked Tiah

Amelia replied, "Well, when we climbed the slide, we were storing up energy just like a ball does when it bounces."

"And as we slide down the stored energy gets changed into energy of motion," Added Tiah.

"And when we reach the bottom of the slide, we run out of stored energy so we had to climb the slide ladder again to store up enough energy to get the ride down," said Amelia.

The girls looked at one another smiling because they really felt they were getting close to solving the mystery of the ball that would not bounce.

"So, the ball must bounce lower because it is losing some of its ability to store up energy, but why?" Asked Tiah, now feeling very puzzled.

"So, now our mystery is to figure out why the ball loses its ability to store up energy," said Amelia feeling a bit puzzled too.

"Wait a minute, I think I know what happened to the energy," replied Tiah excitedly.

"You do, what?" Asked Amelia.

"Remember when we studied energy in science class, Mrs. Tuchman said heat and sound are forms of energy," stated Tiah enthusiastically.

"So," said Amelia

"Well the ball does make a sound when it hits the ground," replied Tiah, "Remember, we did test for that characteristic."

"Brilliant, said Amelia, so some of the energy needed for a bouncing ball changes into sound."

"And sound is not exactly useful when a ball just wants to bounce," added Tiah.

"There you go giving feelings to bouncing balls again," replied Amelia.

"Whatever," replied Tiah.

"Sorry Tiah.

"How do you get heat from a bouncing ball?" Asked Tiah

"Let me think about it a minute. Okay, I think I got it. When a ball touches the ground, some rubbing must take place. And when things rub together, there is friction," said Amelia as she was interrupted by Tiah.

"And where there is friction, heat is released," jumped in Tiah. "Sorry Amelia, but when you started talking about rubbing and friction, I found myself rubbing my hands together and immediately noticed they got warmer. Of course, I already knew that happens when you rub your hands together but I never connected it to a bouncing ball.

"Wow, Tiah, you are really impressing me. See, I said this was going to be fun," said Amelia.

"It has been fun doing this on this rainy day, but we still have not solved the mystery of the ball that does not bounce," noted Tiah.

"But we have solved the mystery of why a ball bounces lower and lower and then stops," said Amelia.

"And that is why? Just kidding," said Tiah. The balls bounced lower and lower and then stopped bouncing because some of the energy needed to bounce was changed into heat and sound after each bounce. Obviously, those are types of energy not very useful to balls that just want to bounce and have fun," continued Tiah.

"Okay, now you got me feeling sorry for those poor balls that can't bounce anymore," joked Amelia.

"And if you think about the slide ride, the rubbing causes friction so heat is released and there is sound as you slide down and when you hit the ground," observed Tiah.

"You are super impressing me, Tiah. Just imagine how the slide ride would go if there was no sound or heat released," remarked Amelia.

"We probably would have broken all of our bones because of going too fast so heat and sound actually saved us a lot of pain," replied Tiah.

"You really need to get more daring," said Amelia, who was known to be a dare devil on the playground.

"Whatever!" Said Tiah.

"Okay, but we still need to figure out why this stupid ball won't bounce," said Amelia, feeling somewhat frustrated.

Amelia was holding a couple of balls in her hands. Tiah picked up some balls too.

"Some of the balls are definitely squishier than others," noted Amelia.

"Well we already know that from our earlier experiment. Remember, I was able to see the squishier a ball is, the more it changes shape when it strikes the ground," Tiah reminded her friend.

"Yes, I remember, but we need to study this some more. Just think, what if a ball cannot change back to its original shape easily? Will this affect the ball's ability to bounce?" Asked Amelia.

"So, we need to do some more testing, but what should we test?" Asked Tiah feeling unsure as to what to do next.

"I got it," Amelia said. "Suppose our hypothesis is that the squishier a ball is, the less it is able to bounce because it does not change back to its original shape. Then when it hits the ground, it exerts less of a force and so the ground also pushes with a lesser force"

"Wow, Amelia, you sure know how to hypnotize," said Tiah

"You mean hypothesize, Tiah not hypnotize," corrected Amelia.

"Whatever," replied Tiah. "So, what do we do next?"

Amelia thought about this. "Well, if I push a balloon against the floor and then let go, it bounces up and goes back to its original shape. But, it doesn't bounce very high." Amelia got up and walked to the kitchen counter. She opened the draw where her mother kept birthday balloons. She brought back three balloons. She inflated one balloon so it was very taut, one medium taut and one was very squishy.

Amelia asked Tiah to bounce each of the balloons while she made observations.

"We already have information about squishy balls," said Tiah.

"I know, but using the balloons will make it easier to observe how the amount of squishiness affects a ball's bounce."

"Whatever," said Tiah.

Tiah first released the balloon with the most air. Amelia noted how many times it bounced. She did the same for the other two balloons.

"Well, no surprise that the least squishy ball bounced the longest," said Tiah.

"Agreed, no surprise," replied Amelia, "but do you know why?'

"It depended on how hard it was able to push on the ground to bounce, just like before" said Tiah.

"Tiah, you will be a scientist yet. That is what I think too. And remember the ground pushes back with an equal force so if the balloon does not return to its full shape, it can't push as hard, so the ground doesn't push as hard too," said Amelia, getting very excited with her insights.

"I am actually understanding what you are saying, Amelia, but how does this explain why the ball we found does not bounce. It doesn't even feel squishy. Did it lose its touch and just could not push anymore?" Asked Tiah, also getting excited with the prospect of solving the mystery.

"Exactly," replied Amelia.

"You are kidding," replied Tiah startled she was actually correct.

"The ball we found lacks the ability to push." Noted Amelia.

"So, the ball stupidly changes all of its energy into heat and sound instead of storing up energy that would allow it to bounce like a regular ball," said Tiah not sure what she was saying actually made any sense.

"Exactly," replied Amelia again.

"I am becoming a scientist. Maybe I will become a rocket scientist," stated Tiah with confidence.

"I wouldn't push it," replied Amelia, but pleased her friend was sharing her interest.

Amelia picked up the ball that would not bounce and dropped it to the floor. It made a thud sound and did not bounce.

"Though this ball looks like an ordinary rubber ball, it is more similar to a very squishy ball that cannot push very hard when it hits the ground. Only, it fooled us, because it did not feel squishy," said Amelia.

"It must be made of something that does not go easily back to its original shape when it hits the ground, only it is not noticeable to us" added Tiah.

Both girls were quiet for a while.

"What are you doing?" asked Tiah.

"Thinking," replied Amelia.

"Oh yeah, I can see the smoke coming from your brain," joked Tiah.

"I think the mystery of the ball that will not bounce has been solved."

"Really?" questioned Tiah

Really," answered Amelia.

"I bet it has something to do with the push of the ball," said Tiah.

"Right, I knew there was a reason we are friends. We both love science." Said Amelia.

"I wouldn't push it that far, but then again I am toying with the idea of becoming a rocket scientist well maybe an assistant rocket scientist," said Tiah with a little sarcasm in her voice.

"We did solve the mystery of the ball that will not bounce. And it is not because it is too lazy to bounce. When it hits the ground, it changes shape slowly so its push on the ground is weak. But just like the ping pong ball and golf ball, it was hard to observe. Then the ground pushes back with the same weak force, but not enough to give the ball a bounce. In the case of our ball that would not bounce, none of its energy of motion could be changed into stored energy. Instead, it all turned into heat and sound," concluded Amelia.

"Who would have ever thought there was so much science that goes into the bouncing of a ball," said Tiah, with an exhausted sigh.

"Look, it stopped raining, let's go outside and play soccer in my yard," suggested Amelia.

"Brilliant, our brains can use a rest for now and you know what they say, 'a strong body makes for a strong mind'," replied Tiah, happy to run around outside, but feeling pleased to have solved the mystery of the ball that would not bounce with her best friend, Amelia.

"Hey Amelia, maybe next time in science class, I can help you."

"Whatever," replied Amelia.

Discussion Questions

1. How could Amelia and Tiah improve their experiment when they dropped the balls?
2. Why do some balls not bounce as high as other balls?
3. Why can't a ball bounce forever?
4. Using Newton's three Laws of Motion, explain the movement of a bouncing ball.
5. The ball that did not bounce was made of a certain type of matter. Can you think of any other uses for this type of matter?
6. What types of energy changes did Tiah and Amelia observe watching a ball bounce?

The Case of the Ball That Would Not Bounce Science Terms

1. **Gravity:** A property of matter that has an attraction for other matter. The more matter, the greater the attraction.
2. **Force:** A push or pull
3. **Friction:** A force that opposes motion.
4. **Gravitational Potential Energy:** Stored energy dependent on the weight of matter (from pull of gravity) and the height from which it falls. Mathematically expressed as: GPE = $9.8 ms^2$ x mass x height.
5. **Heat Energy:** Energy that moves from an area of higher temperature to one of lower.
6. **Hypothesis:** A prediction based on observations.
7. **Kinetic Energy:** Energy of motion. Mathematically expressed as $1/2M \times V^2$
8. **Motion:** Any movement or change in position relative to an observer.
9. **Newton's Third Law: Action Reaction:** For every action, there is an equal but opposite reaction
10. **Potential Energy:** Stored energy
11. **Sound Energy:** Energy produced by the vibrations of matter.
12. **Temperature:** A measurement of the average kinetic energy in matter.

NGSS

MS-PS1 Matter and Its Interactions
Disciplinary Core ideas
Definitions of Energy
- The term "heat" as used in everyday language refers both to thermal motion (the motion of atoms or molecules within a substance) and radiation (particularly infrared and light). In science, heat is used only for this second meaning; it refers to energy transferred when two objects or systems are at different temperatures.

Developing Possible Solutions
- A solution needs to be tested, and then modified on the basis of the test results, in order to improve it.

MS-PS2 Motion and Stability: Forces and Interactions
Disciplinary Core Ideas
Forces and Motion
1. For any pair of interacting objects, the force exerted by the first object on the second object is equal in strength to the force that the second object exerts on the first, but in the opposite direction (Newton's third law).
2. The motion of an object is determined by the sum of the forces acting on it; if the total force on the object is not zero, its motion will change. The greater the mass of the object, the greater the force needed to achieve the same change in motion. For any given object, a larger force causes a larger change in motion.

Types of Interactions
1. Gravitational forces are always attractive. There is a gravitational force between any two masses, but it s very small except when one or both of the objects have large mass—e.g., Earth and the sun.

MS-PS3 Energy
Disciplinary Core Ideas
Definitions of Energy
1. Motion energy is properly called kinetic energy; it is proportional to the mass of the moving object and grows with the square of its speed.
2. A system of objects may also contain stored (potential) energy, depending on their relative positions.

Relationship Between Energy and Forces

1. When two objects interact, each one exerts a force on the other that can cause energy to be transferred to or from the object.

Conservation of Energy and Energy Transfer

1. When the motion energy of an object changes, there is inevitably some other change in energy at the same time.

Science and Engineering Practices
- Constructing explanations and designing solutions
- Obtaining, evaluating and communicating information
- Planning and carrying our investigations
- Engaging in argument from evidence

Crosscutting Concepts
- Patterns
- Cause and effect
- Scale, proportion and quantity
- Energy and matter

The Case of the Dent on the Roof of Mom's Car

Walter and Evie were walking to school together. "Wow, mom was on the warpath this morning," said Walter.

"She is very upset about the big dent on her car's roof" replied Evie to her twin brother, Walter.

"Dad wasn't too pleased himself," added Walter.

Evie and Walter's mother and father could not understand how such a large dent formed on the roof of the car. Their dad discovered the dent while he was taking out the trash. He is pretty tall and as he passed the car, he noticed the dent on the roof.

"Hey, remember when the deer ran into mom's car door last summer?" said Walter. "Yeah.' Replied Evie, "mom thought that there would be a big dent, but there was only a little scratch."

Though Evie, Walter and their mom were startled by the big thud they heard when the deer hit the door, it did not hit it very hard, and just continued its journey across the road, apparently, unhurt.

"What could have caused such a big dent on the car's roof?" They remembered their dad asking over breakfast that morning.

The twins decided to take on the challenge and figure out the cause of the dent on the car's roof. Evie and Walter felt that if they could determine what was the cause of the dent, it might help prevent it from occurring again. They also thought it would be fun to try to solve the mystery of the car's dent. They loved mysteries and were pretty inquisitive twins! In addition, they felt bad their parents were upset about the cost of repairing the dent. After listening in on their parents, they learned how insurance companies have deductibles and the deductible on the damage to the car was still pretty high so their parents had to pay a lot of money out of pocket.

When Evie and Walter got home from school, they decided to inspect the dent and gather some information about it. They obtained a step stool and each took a turn climbing up to take a closer look at the dent.

"Hey Walter did you notice that some of the paint in the dent is chipped?"

"Yeah, I did," replied Evie "Something hit the roof with enough force to also damage the paint. What in the world could it had been?"

"I'll get my notepad," continued Evie. "We need to write everything down."

"You can write everything down," replied Walter. "I do enough writing-down in school."

"Based on what I have seen, you can use the practice," teased his twin.

"The only practice I need is in baseball," said Walter getting annoyed at his sister.

"Okay, okay, calm down, Walter. I'll take all the notes," replied Evie.

"Good, now go get a ruler," ordered Walter to his sister.

"And why do we need a ruler?" Asked Evie.

"To measure the size of the dent," replied Walter.

"Right! Sorry for making fun of your writing," said Evie.

"Apology accepted," replied Walter.

Evie went to her room to get the ruler she uses for her science class.

When she returned, she climbed the step stool to begin the measuring.

"Measure in metric," said Walter.

"On it," replied Evie.

Walter suddenly realized he was holding the notepad when his sister went to get the ruler.

"Wait a second, Evie, here's the notepad and pencil so you can write down the measurements," said Walter.

"Walter, can't you do it. It is kind of tricky measuring and writing at the same time while on a step stool," said Evie.

"I told you I did not want to do the writing so you come down the ladder and I will do the measuring and you do the recording," responded Walter.

"You really have to make things more difficult than they are, but okay," said Evie.

She climbed down the ladder and gave the ruler to her brother and he gave her the notepad.

Walter measured the length, width and depth of the dent. He called to his sister, "The dent is 0.5 cm deep, 5 cm wide across the smallest section of the dent and 8 cm wide across its largest section. It's not huge, but definitely noticeable and needing repair," he continued.

"Mom reminded dad that if it was not repaired there would be some rusting and the damage would get worst," said Evie.

"Mom and dad like everything to be tidy," said Walter.

"Right, that is why they are always on our case to keep our rooms neat. I think when I become a parent, I will give my kids more slack. A messy room shows a busy child," rationalized Evie.

"Evie, let's get back to our investigation and stop worrying about tidiness," said Walter trying to get his sister back on task.

"Well since we now know the size of the dent, what does this tell us?" said Evie, thinking out loud.

"It tells us that whatever hit mom's car had enough force to make a good size dent," said Walter "So what type of object exerts enough force to cause a dent like this," he continued.

"I have an idea, why don't we test out different objects to determine what factors affect the force an object has on another object," suggested Evie.

The twins reasoned that the data they collected could be applied to what hit their parent's car. Evie suggestion gave Walter an idea.

"Hey Evie, remember in science class when Mrs. Dayton used a tub of flour to test the impact of a meteorite. We dropped different size rocks into the flour. Well, we can do the same thing and drop different objects into flour to see what factors affect the force an object has on another," suggested Walter enthusiastically.

"I don't think mom will be too happy with us using the flour, but we do have some bags of sand in the garage and that should work pretty well," said Evie.

"Okay, I'll get the sand and you find a container to put it in," said Walter. "It was fun making those craters in the flour during Mrs. Dayton's class."

Evie found a large plastic storage container. They went around their house and gathered together objects of different weight and size. They decided to do the experiment in the basement of their house so not to make a mess in their house knowing how tidy their parents are. They both worked and coming home to sand all over the kitchen floor would not be a good welcome home for them.

The twins brought all of their supplies for their experiment down to the basement.

"Walter, you are putting too much sand in the container. You don't want it to overflow," warned Evie.

"Check," replied Walter and he removed some of the sand so it was not overflowing out of the container.

Evie and Walter looked over the objects they had gathered.

"We sure have a good selection of objects, said Evie.

Their objects for testing are a ping-pong ball, golf ball, softball, a 16 oz can of unopened tomato sauce, an inflated balloon (diameter= 30 cm), quarter and rock (diameter = 12.5 cm).

"We need to weigh each object to see how that affects impact," said Walter.

"We can use the kitchen scale mom uses to weigh food," suggested Evie.

"Why does mom weigh her food?" Asked Walter.

"You know Walter, it's when mom goes on one of her crazy diets," said Evie.

"Oh yeah! Mom is hard to live with when she goes into the 'diet zone'," said Walter.

Evie went up to the kitchen and brought the scale down to the basement.

"Well now her scale is finally serving a very important purpose," said Evie.

"Should we use ounces or grams?" Asked Walter.

"Let's use grams like what we do in science class," answered Evie.

"OK, Evie but you should say we are massing the objects not weighing them. You know Mrs. Dayton would go bonkers if you mix up weight and mass. I will do the massing and you record down the mass on your notepad.

"Check," replied Evie as she picked up her notepad.

Walter massed each object and Evie recorded down the measurement.

"Now the fun part begins," said Walter. "I get to drop the objects into the sand.

"And measure their impact," added Evie.

"We also need a tape measurer, Walter"

"Why do we need a tape measurer?"

"Well to be scientific, all of the object should be dropped from the same height so we need to measure the height. Our metric ruler is only 30 cm long so that would not be very high," explained Evie.

"Good point, Sis"

"Come to think of it, we should drop each object from different heights to see how distance affects the size of the crater the object makes in the sand like what we did in science class with the rocks," suggested Evie.

"Let's drop the objects from 100 cm, 75 cm, 50 cm and 25 cm. This way we don't have to use the step stool, which is still in the garage" said Walter

"OK and then after each drop, you measure the length, width and depth of each dropped object. I will make a chart to record down the results," said Evie.

"And to be really scientific, we should repeat each drop to check to ensure accuracy of drop," Added Walter.

"How many times 2, 3, 4?" Asked Evie.

"Let's do it 3 times and take the average. That should help in accuracy," replied Walter.

"You know, maybe we can be hired by insurance companies to investigate the cause of car dents," continued Walter.

"You are always thinking of angles to make money. I doubt there would be interest from insurance companies. However, mom and dad would be interested in how the dent happened since it did occur in a weird spot. We know it wasn't a fender bender," said Evie.

"Well some day, one of my ideas will be a money maker, you will see. Anyway, though mom and dad are not going to pay us. I agree, they will be happy to know the cause of the dent and what brilliant twins they have," said Walter.

"As your twin, I will agree with the brilliant part," responded Evie.

"Enough talking, let's begin the experiment since we have everything we need now," said Walter.

"Actually, we don't," responded Evie.

"What do you mean?" Asked Walter.

"Well if we are going to do an experiment, we need a hypothesis," answered Evie.

"We are not in science class. We are not getting a grade on this experiment," replied Walter wanting to just get on with the experiment.

"No, we need a hypothesis. We cannot do an experiment without a hypothesis," said Evie.

"Though you are my twin, you are really weird, Evie."

Evie stared back at her brother.

"Okay, okay, we will formulate a hypothesis like we were in a science class," said Walter.

"What's your hypothesis?" Asked Evie.

"No, what's yours," countered Walter.

"Well, we should hypothesize about how the mass of an objects affects the size of the crater it makes," said Evie.

"And we should include how the height from which it is dropped affects the size of the crater," added Walter.

"I remember when we did the meteorite experiment with rocks, the heavier rocks formed deeper craters so my hypothesis is the greater the mass of an object, the deeper the crater," said Evie.

"I am not positive how height will affect the depth of a crater, but the higher an object is dropped, the faster it will probably be going when it hits the sand so my hypothesis is objects dropped from higher points would produce deeper craters because they would be moving at higher velocities than objects dropped from a lower height," said Walter.

"Great, I wrote down our hypotheses and we are finally ready to begin the experiment," said Walter.

Walter dropped each object four times to correspond with the four heights. He measured the length, width and depth of each crater. He repeated the drop for each object two mores times for a total of three trials as they decided for a more accurate result. They used the average measurements for each distance dropped.

Evie carefully recorded down all measurements on the chart she generated.

When Walter looked at the chart, he said, "Evie, I cannot read your handwriting. Why is it so sloppy?" He asked.

"I had to write fast because you were calling out a lot of information," Evie replied.

After looking at her handwriting, Evie said, "I guess you are right, this is really hard to read. Now I can appreciate how hard it is for teachers to read all the crazy hand writings in a classroom."

"Evie, let's place all the data on an excel document. I know how to set up an excel chart," said Walter.

"I can set up an excel sheet too," said Evie thinking her brother was trying to show off.

"I am not questioning that you can generate one too, but since I cannot read your handwriting, I will set up the excel document and then you read me the data we collected."

Evie knew her brother was right and apologized for accusing him of showing off his talent as if he was the talented twin.

The twins worked together generating an excel chart with all of the data they had collected.

When it was completed they both carefully looked over the data.

"I bet the crater left by the object's impact would have been deeper and cleaner if we used flour, but it was not worth the wrath of mom," noted Walter.

"We still have interesting results," said Evie.

"Yes, we do," agreed Walter.

Object	Mass (g)	Average Width (cm)	Average Length (cm)	Average Depth (25 cm)	Average Depth (50 cm)	Average Depth (75 cm)	Average Depth (100 cm)
Ping pong ball	2.3	4	4	.3	.5	.7	.9
Golf ball	45.9	4	4	.8	1	1.3	1.4
Softball	177	10	10	1	1.5	1.8	2
Can	280	8	15	2	2.5	3	3.3
Balloon	1.55	15	16	.2	.5	.5	.5
Quarter	5.67	2	2	.3	.5	.7	.8
Rock	75	4	6	.5	.7	1	1.2

"What do you notice?" Asked Evie.

"Well you don't have to be a rocket scientist to see that the mass and size of an object affects the size of the crater it creates when dropped," said Walter.

"Walter it looks like your hypothesis that the higher an object is dropped the deeper the crater is correct," noted Evie.

"As I said before, it is because when an object is dropped, the higher up it is, the longer it is falling. The longer it falls, the faster it is going on impact" noted Walter.

"Now I see, Walter, if two cars collide, the faster they are going, the greater the damage to the cars. So, velocity does have something to do with force in addition to the mass of an object," said Evie.

"Actually, I am not surprised and you should not be either," replied Walter.

"I have a feeling this has something to do with Mrs. Dayton's science class."

"Evie, can you say Newton?" Asked Walter.

The Case of the Dent in Mom's Car, <u>Tales of Science</u> by Joan S. Wagner

"As in Isaac? Oh yeah, now I remember. This all has to do with Newton's Laws of Motion," said Evie.

"Right, it was what we studied last month in science class." Said Walter.

"Well, Newton has 3 laws of motion," stated Evie in a matter of fact manner.

"Do you know which law applies to our results? Asked Walter.

"It's the one that has to do with force," replied Evie. I remember it was sort of like a formula of some type," said Evie, wishing she could remember the formula before her brother did.

"It is Newton's Second Law of Motion. The amount of force created by an object is directly related to its mass and its final velocity," stated Walter. "When objects fall, they accelerate so the longer they fall, the faster they go, the greater its force" he continued.

As he was talking, Evie had a big grin on her face.

"Why are you smiling?" Walter asked.

Evie held up the piece of paper.

On it was the formula showing Newton's Second Law of Motion.

Force = Mass x acceleration or F=ma

"You are my twin," smiled Walter back. "I figured you would remember it."

"I guess you have more confidence in me than I am, but thanks for the support," replied Evie. "But, I am still confused about something," she continued

"What?" Asked Walter.

"Why does a falling object accelerate?" Asked Evie feeling stumped.

"Why do you think things fall?" Walter asked.

"Because of gravity," answered Evie.

"Right!" said Walter.

"Sooooooo!" Said Evie.

"Remember when Mrs. Dayton dropped the textbook and a pen and they both fell at the same time to the ground?" Asked Walter.

"Yes, I remember," replied Evie.

"Do you remember Mrs. Dayton's explanation? Asked Walter of his twin trying to coax her into an answer.

Suddenly, a big grin came across Evie's face. Walter recognized the grin.

"So, explain, Evie," said her brother

"Gravity accelerates objects as they fall so the longer they fall, the faster they go," said Evie remembering the science lesson.

"Do you remember what the acceleration of gravity is?" Asked Walter.

"No, but I can Google it," replied Evie.

"I will save you the time. Gravity accelerates 9.8 meters a second every second, said Walter.

"Oh, I remember so for every second an object is dropped its velocity increases by 9.8 meters a second. So, the faster an object moves, the greater its force," said Evie.

"Wow, you are impressing me," said her brother.

"Now you got me thinking," continued her brother, "remember when our science teacher told us how NASA was puzzled by pock marks on the windows of the first space shuttles. When the shuttles were brought home, it turned out that paint chips from old satellites were the culprits. Though the paint chips have little mass, they are moving at high velocities. So, when the shuttle collides with speeding paint chips, the force of impact caused the pockmarks in the shuttle's windows."

"Walter you may think this is weird but I actually remember the speed of the paint chips. They travel at about 17,500 miles an hour, the same as the shuttle," said Evie.

"So, you can remember that number but not the acceleration of gravity," said Walter.

"I think it is because it was weird to think a paint chip can cause that type of damage," replied Evie.

"Though you are my twin, you can be pretty strange," said Walter.

As Walter was talking, Evie was observing the excel chart again and said, "Look at the results for the balloon, the depth of its impact did not seem to change."

"Right, Evie, let's drop it a fourth time and observe more closely what happens as the balloon falls and hits the sand from each height," said Walter.

As the balloon dropped, Walter said, "look, the balloon does not fall the same way as the other objects. It seems to float down."

"I don't know why I did not pay attention to that before," said Evie.

Though Evie may not remember from her science class, the balloon floats down because it does not exert as much of a downward force as the other objects so air pressure makes it appear to float down.

"Well now we know it was not a balloon that caused the dent in mom's car, said Walter, jokingly.

"Very funny, bro."

Evie and Walter gave the excel chart a final review.

"So, let's summarize what we know," said Evie.

"We do know that the greater the mass and the faster an object moves, the greater its force," said Walter.

"And the pock marks cause by paint chips is certainly a good example of how velocity affects force," said Evie.

"Do you know how they determined it was paint chips, Evie?"

"Yes, the reason they knew it was paint is that the paint was in the pock mark." replied Evie.

"That was a cool discovery," said Walter.

"You know NASA is very concerned about all the space junk colliding with shuttles and satellites. Just like our planet, space is going to need a clean up or there is going to be more and more reports of space junk colliding with shuttles and satellites or even space telescopes," said Evie.

"Well we won't be surprised if a big news story tells about space junk destroying a communication satellite," said Walter.

"There will be a lot of unhappy people if they cannot get cell service or programs on their TVs," added Evie.

"Let's get back to solving the mystery of the dent in mom's car," said Walter.

"Well based on the data and shape of the dent, it looks like it was caused by a ball, perhaps a golf ball, said Evie.

"How in the heck did a golf ball hit mom's car? Asked Walter. "Mom doesn't play golf"

"I think it is time to interview mom to gather some more critical data," said Evie.

After talking to their mom, they learned that she had lunch by a public golf course. Evie and Walter reasoned that someone hit a golf ball that veered off the course and landed on the car's roof causing the damage. It must have been a fairly high hit in order to fall on the car and cause the dent, they concluded.

"Yes," said Walter. "The greater the motion and mass of an object, the greater the force that can be generated."

"Sir Isaac Newton must have been a very smart person to figure out the laws that affect the motion of objects," said Evie "And to think that he developed these laws over 300 years ago," she continued.

"Did you know that he invented a refracting telescope and even figured out that white light contains all the colors of the rainbow," said Walter who was on the computer reading about Newton in Wikipedia.

"Mrs. Dayton said he invented calculus," added Evie.

"I think the next time we are assigned to write a report about a famous person, I am going to choose Sir Isaac Newton," said Walter.

"Newton is interesting, but I think I rather do a report about a woman scientist like Madam Curie," said Evie.

The twins sat down together, exhausted and hungry. They thought about their accomplishment.

"You know doing science is like being a detective," said Walter. "You make observations, you look at the clues, gather data and voila, you reach some conclusion."

"However," added Evie, "sometimes your conclusion may turn out to be wrong or in need of some modification."

"Right, I remember Mrs. Dayton saying we learn more when a hypothesis turns out to be false because it causes us to rethink the problem," noted Walter.

Evie and Walter realized that new information might be discovered that may show them that it was not a golf ball, but something similar to one that caused the dent. If the new ball had some of the car's paint on it, like the paint in the pock marks of the space shuttle, the chance that the ball with the car's paint on it caused the dent was close to 100%. It was unequivocal. But, alas, they did not have the golf ball that hit their mom's car.

The Case of the Dent in Mom's Car Discussion Questions

1. What does mass and velocity have to do with the dent in the car?
2. Why couldn't the twins state unequivocally that the dent was caused by the golf ball?
3. What was the purpose of testing other objects to see the size of a dent?
4. Why was it useful to create a chart of the objects dropped?
5. How would you improve Evie and Walter's investigation?
6. Try out Evie and Walter's investigation with your improvements. Here is what you will need:

    Metric ruler
    Assorted balls: ping pong, golf, softball
    Rock
    Quarter
    16 oz can of tomato sauce
    balloon

The Case of the Ball That Would Not Bounce Science Terms

1. ***Force****:* A push or pull.
2. ***Friction****:* A force that opposes motion.
3. ***Gravity****:* A property of matter that has an attraction for other matter. The more matter and the closer together the matter, the greater the attraction. Mathematically, Newton described it with this formula: $G = M_a \times M_b/d^2$
4. ***Gravitational Potential Energy***: Stored energy dependent on the weight of matter (from pull of gravity) and the height from which it falls. Mathematically expressed as: $GPE = 9.8 ms^2 \times mass \times height$.
5. ***Heat Energy***: Energy that is transferred from an area of higher temperature to one of lower.
6. ***Kinetic Energy****:* Energy of motion. Mathematically expressed as $1/2 M \times V^2$.
7. ***Motion****:* Any movement or change in position relative to an observer.
8. ***Potential Energy***: Stored energy.
9. ***Sound Energy****:* Energy produced by the vibrations of matter.
10. ***Temperature****:* A measurement of the average kinetic energy in matter.

NGSS Standards

MS-PS2 Motion and Stability: Forces and Interactions
Forces and Motion
1. The motion of an object is determined by the sum of the forces acting on it; if the total force on the object is not zero, its motion will change. The greater the mass of the object, the greater the force needed to achieve the same change in motion. For any given object, a larger force causes a larger change in motion.
2. All positions of objects and the directions of forces and motions must be described in an arbitrarily chosen reference frame and arbitrarily chosen units of size. In order to share information with other people, these choices must also be shared.
3. Gravitational forces are always attractive. There is a gravitational force between any two masses, but it s very small except when one or both of the objects have large mass—e.g., Earth and the sun.
4. Plan an investigation to provide evidence that the change in an object's motion depends on the sum of the forces on the object and the mass of the object.

Science and Engineering Practices

1. Designing and using models
2. Analyzing and interpreting data
3. Constructing Explanations and designing solutions
4. Obtaining, gathering and communicating information

Crosscutting Concepts

1. Cause and Effect
2. Scale, proportion and quantity

Staying Alive

It's All in the Structure

Lauren, Gavin, Grant, Hillary and Brady were walking to the park together. Hillary was baby-sitting for her little sister, Olive who was dragging her stuffed beagle along.

"Stop dragging your beagle. You will hurt it," said Hillary to her little sister, who continued to drag it.

"Come on Hillary, how can you hurt a stuffed animal. It's not alive," said Grant.

"Oh yeah! I bet you can't prove it," retorted Hillary.

"Just try me," replied Grant looking forward to a little challenge.

"This should be interesting," whispered Gavin to Lauren.

"Living things can move," challenged Grant to Hillary.

Hillary pulled the beagle from her startled sister's arms and wound it up so it began to move. "There, it is moving," said Hillary.

"You have to be kidding," retorted Grant. "Living things can move on their own. They do not have to be wound up."

"It looks like a point for Grant," chuckled Gavin to Lauren.

"Well living things need to eat. They take in food," said Hillary trying to show Grant what it means to be alive.

However, Grant was determined to challenge Hillary so he said, as a car passed them by, "Cars eat. They eat petroleum."

Hillary realized that she would have Grant on this one. She replied, "True, but cars use the petroleum only for energy. The petroleum is not converted into more car or to repair damaged car parts. Living things use food for energy, growth and to repair damaged parts.

"It looks like the score is tied," said Gavin to Lauren, grinning.

Grant next took a piece of paper and tore it into smaller pieces, which he placed into a trash bin they passed on the way to the park. He then declared, "See Hillary, the paper just reproduced."

Gavin and Laura waited to see Hillary's next response. Little Olive continued to suck her thumb.

"True, you made more paper," Hillary responded. "However, the paper cannot take in food and grow into a big piece of paper because it is not alive." Hillary then waved her hand in front of Gavin's eyes and he blinked. "See," said Hillary, only living things can respond to a stimulus."

Grant thought a minute and then said, "when I send a signal to my remote-control car, it responds by moving. When I push a key on my computer, it responds to what the key asks it to do."

Grant knew that remote control cars and computers were not alive, but he liked to see how his friend, Hillary would respond. The two of them often "played" games like this.

Gavin and Lauren waited for Hillary's response.

Hillary realized that Grant was 'playing' with her again, but it was fun to think about what made something alive.

Finally, Hillary replied, remembering what she had learned in science class. "OK, Grant, there are many things that non-living things appear to do, but there are some major differences. All living things are composed of one or more cells. So, when living things grow, they make more cells. Non-living things cannot do that. Though they get rid of wastes through the tailpipe, it is not the same as the wastes excreted by living things."

"How are they different challenged," Grant.

Hillary responded, "The wastes excreted by living things come from life activities carried out by all of our cells. Food that is not digested cannot get to our cells so is rid of or egested, to use a science term, as feces. So, on the surface it may seem like a car is alive because it takes in 'food' and excretes wastes, but that does not come from cellular activity."

"OK, you win," said Grant. "There are many ways by which living and nonliving things differ. And yes, though cars excrete wastes, I know it is due to the actual burning of the fuel. However, you do have to admit that cars appear to be alive because they move, take in "food" and release wastes. They even respond to the environment."

Gavin was puzzled when Grant said cars respond to the environment so he questioned him on it.

Grant answered, "Well if you push on the breaks, a car slows down or stops.

Lauren decided to get into the discussion. She said. "The car slows down because a signal is sent by a computer to the brakes."

Though Grant knew that a car was not alive, he continued to challenge his friends. "Your brain is like a computer. If you are running and want to slow down, your brain sends that information to your body and it responds by slowing down."

Hillary was listening to the discussion and though she realized Grant was 'playing' with all of them, she decided to bring the discussion to a conclusion. "On the surface, many non-living things may appear to be alive, but since they are not composed of cells, they cannot be alive and that is the main difference" she said.

While at the park, Hillary took Olive to the swings while the others played ball. After pushing Olive on the swings, she walked over to the pond with Olive. Hillary thought it would be fun to look at what is in the pond water. She had gotten a microscope for her birthday the year before. She emptied her water bottle and filled it with pond water. She knew her mother would be upset that she used the bottle, but it was the only container she had.

After the park, they all went back to Hillary's house. Her mom was back from her errands. She thanked Hillary for babysitting and put Olive down for a nap. Olive was exhausted from the park and all the talk of living vs. nonliving. It made her head spin, since she was not quite sure what they were arguing about.

Lauren, Gavin and Grant asked Hillary what she wanted to do since they were at her house. Hillary always thought of herself as a budding scientist, and since they already had discussed what it is to be alive she said. "While we were in the park, I took a sample of the pond water. I thought it would be fun to see what is in the pond water. In fact, I can see some things swimming around, but it is hard to see what they actually look like. I got a microscope for my birthday last year so I thought we could look at the pond water under the microscope."

"We did that in science class," said Lauren. It was fun. The teacher had asked all of us to bring in pond water samples.

Hillary went to the closet to get her microscope kit. It came with some prepared slides, plain slides and cover slips, iodine and pipettes. She placed the microscope carefully on the table. "We can take turns looking through the microscope," she said to her friends. She prepared four slides by placing a drop of the pond water in the center of each slide and carefully dropped a coverslip on it so there would not be any air bubbles

She said, "You have to be careful not to create air bubbles because they block what you are trying to see and you could mistake them for something alive because they do move around on the slide."

Lauren remembered having that problem when she looked through her microscope in Mrs. Squires' class. "I thought a water bubble was a single-celled organism swimming across my slide. Mrs. Squires said that many students often make that mistake."

Grant said, "I can see something swimming on this slide."

Hillary's mom walked in and gave Grant a magnifying lens. "Here, this should help you see it a bit better until you can view it under the microscope," she said.

"It is probably a Daphnia," said Lauren. "It is a water flea that is pretty common in ponds. You can see its heart beating easily. In science class we did an experiment to see if temperature affects how often the heart beats."

"What was your prediction?" Asked Gavin.

Lauren replied, "I hypothesized that the cold would cause the heart to slow down because this is true of animals that hibernate in the winter. It turned out that the heartbeat did slow down at lower temperatures and speeded up at higher temperatures. It was interesting and fun doing the experiment"

"Why don't we try repeating Lauren's experiment," said Gavin.

"We would need a lot of Daphnia and I doubt we have many in the pond water I sampled," replied Hillary.

Hillary looked through her microscope and sure enough she spotted a Daphnia on her slide. "They are neat looking and you can really see their heart beating."

"How many beats does it have in a minute?" asked Gavin.

"Shh, I am counting," replied Hillary. "OK, there are 90 beats in 30 seconds so I would say about 180 beats a minute if the beats are regular."

She repeated counting but did it for a minute and got 178 beats. She found this tricky and tiring since the heart beats so quickly. "I guess you can estimate the number of beats in minutes by counting the beats in 30 seconds because the Daphnia's heart beats in a regular rhythm."

"Perhaps you just have to count the beats for 15 seconds and multiply by 4 to get the number of beats per minute," suggested Lauren.

Hillary tried it and after multiplying by 4 got 186. "I guess we can do it that way too," she said.

Hillary's friends each took a turn looking at the Daphnia and counting heartbeats.

When Grant looked through the microscope he described a cell that kept changing shape. "It looks like the blob from a science fiction movie I once saw," said Grant.

"Oh, I bet you are looking at an amoeba," said Lauren. Her science class had looked at a lot of pond water organisms. "Amoeba are unicellular organisms while Daphnia are multicellular ones."

"I see a circular structure inside of the amoeba. Do you know what that is?" Grant asked Lauren.

"Let me take a peek," said Lauren. "Oh, you are looking at the nucleus of the cell. It is sort of like the brain of the cell or in this case, the brain of the amoeba."

Hillary went to the computer and printed out a picture of the Daphnia and the amoeba. She called her friends over to take a look at the pictures.

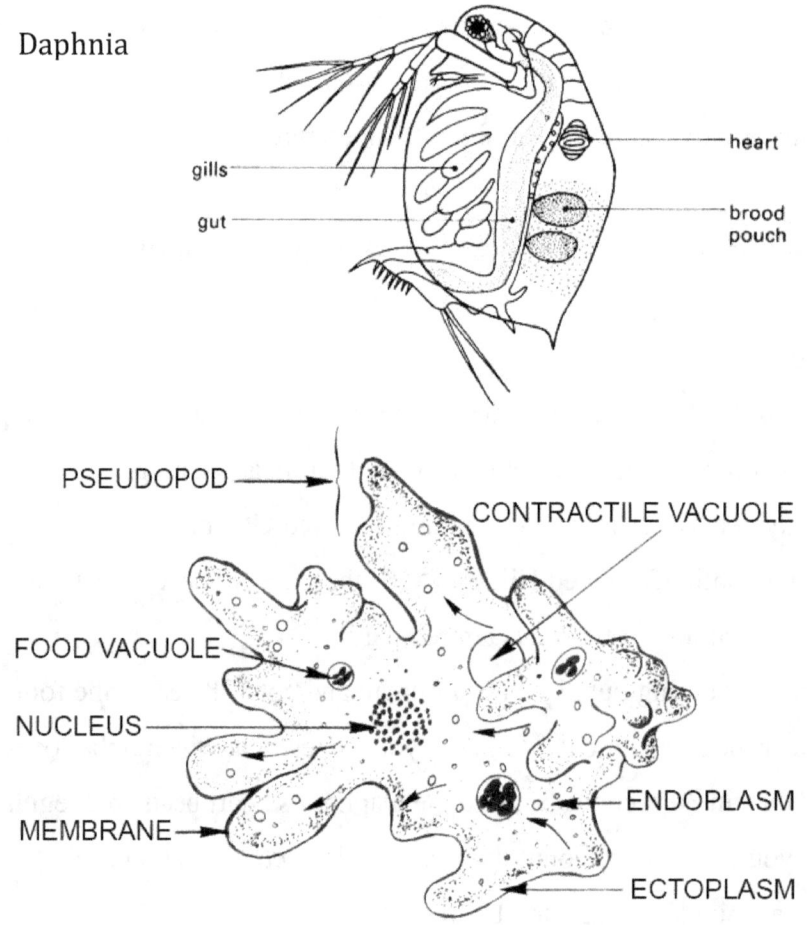

"Wow, it's hard to believe the amoeba has so many parts in it and it is only one cell," noted Gavin.

"Look the amoeba has a food vacuole. It is sort of like our stomach because it can store and digest food," stated Hillary.

"What are the pseudopods?" Asked Gavin.

Hillary replied, "*Psuedo* means fake and *pod* means foot, anyway that is what my science teacher told us."

"So, Amoebae have false feet, sort of like hands to take in food," said Lauren as she looked over the pictures, pleased she remembered this from her science class. "As I said before, the nucleus," she pointed to a structure in the diagram of the amoeba, "is sort of like its brain. My teacher calls all these parts inside of a cell, organelles because they are like little organs."

"What is the cytoplasm?" Asked Grant.

"That is a jelly-like substance in which all of the organelles are located. It is held together by the cell membrane," read Gavin reading from Wikipedia.

"It looks like I have here a bunch of budding scientists," said Hillary's mom as she walked by.

Lauren looked up at Hillary's mom and smiled. She then elaborated on what Gavin had said. "The cell membrane is very important. My teacher said that the cell membrane only allows certain substances to go in and out of a cell."

"Right," said Hillary, "or else the cell would always be leaking if everything can pass in and out of it.

Gavin was looking through the microscope and saw some green cells. "What are these green cells?" Asked Gavin.

After Grant took a look, he said, "I will Google green cells in pond water and see if I can find the answer."

Hillary knew that the green was most likely chlorophyll, which plant-like organisms use to make food. "Those are probably the chloroplasts in the algae cells on the slide. Chloroplasts are the organelles that hold chlorophyll," she stated.

After researching on the computer, Grant said, "Photosynthesis is the life process by which plants make their food using the chlorophyll in the chloroplasts. Using energy from the sun captured by the chlorophyll, water and carbon dioxide combine to form simple sugars which the cell uses for food."

"So, chloroplasts are like food factories." said Gavin.

"Exactly," agreed Grant.

"Hmm," said Hillary after looking at some algae pictures on the Internet, I think we are looking at two types of green algae. One is called Chlamydomonas. They can move around with flagella and even have an eyespot that is sensitive to light, but they are very small so hard to see its features. The other algae grow into filaments. It is called Spirogyra because the chloroplasts appear to take on a spiral shape. Take a look at these pictures. They look like what we have on the slide."

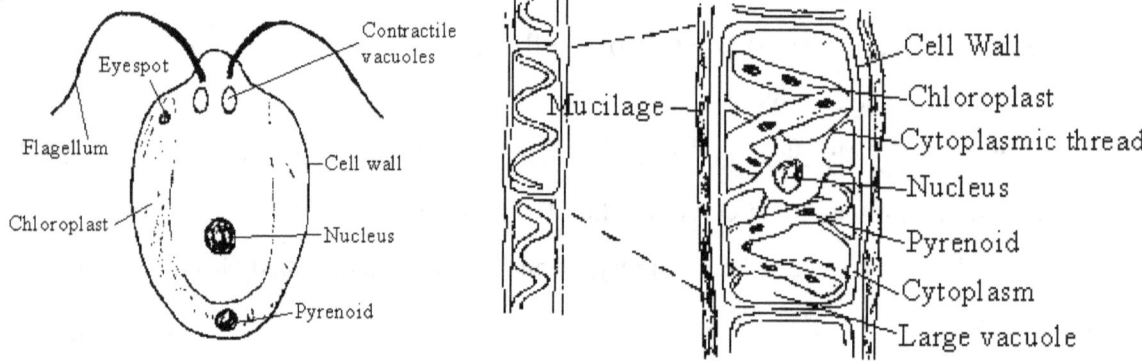

"I think most of us already know that," interrupted Lauren.

They all took a turn looking through the microscope and agreed with Hillary.

Gavin noticed that a plant cell has a cell wall while an animal cell does not. "Plant cells seem to have a more rectangular shape," said Gavin.

"it is because of the cell wall. Anyway, that is what Mrs. Squires, my science teacher said.

"Makes sense to me. What do you want to do now?" Asked Gavin.

Lauren thought it would be fun to compare and contrast the Amoeba and the Daphnia. Her science teacher often challenges the class to do that. As she was thinking this, Grant said, "Wow, the Amoeba and Daphnia sure do look different from one another."

Lauren used Grant's statement to suggest they try to list all the ways in which they are different and then ways in which they are alike.

"Well, they are both alive," stated Grant.

"They both need to breathe," said Hillary.

"They both respond to stimuli in the environment," added Gavin. "They both take in food," said Lauren.

"They both need energy to carry out all of their life activities," stated Hillary.

"Where is the energy produced?" Asked Gavin.

"Time for Wikipedia," said Lauren.

Lauren went to the computer and wrote in the Google search, "What structure in the cell produces energy?" She found the website titled 'Biology4kids.com' most useful at the URL: http://www.biology4kids.com/files/cell_mito.html. After she read over the website she told her friends, "Structures in the cell called the mitochondria are where energy is produced. They are very small so you need a high-powered lens on the microscope to see them."

"So, Daphnia and Amoeba both have mitochondria," said Gavin.

Once she heard the word, mitochondria, Lauren realized it was discussed in her science class. "I remember my teacher saying that the number of mitochondria in a cell depends on the energy needs of the cell," said Lauren.

"Then muscle cells must need a lot more mitochondria than skin cells since they move a lot," concluded Grant.

They all agreed with Grant that it makes sense for muscle cells to have more mitochondria.

Getting back to how Daphnia and amoeba are alike, Hillary said, "Another similarity is that they both can reproduce."

"And just like we need to get rid of our garbage, both have to get rid of wastes," added Lauren.

After gathering similarities between Daphnia and Amoeba, the friends began a list of how they are dissimilar.

"Daphnia are multicellular and Amoeba are unicellular," stated Grant.

Since Lauren had studied some of this in her science class, she was a little more knowledgeable than her friends. She said, "Daphnia's cells are organized into systems to carry out life activities such as digestion and excretion of wastes. Since Amoeba are unicellular, they have no systems."

"I know another difference," said Hillary, remembering what she learned in science class. "Daphnia reproduce with eggs, Amoeba divide in half, called binary fission."

"According to the diagram, Daphnia have gills to breathe, Amoeba must breathe through their cell membrane," said Grant.

"Do you all know the difference between breathing and respiration?" Asked Lauren of her friends.

"Sure, said Gavin. Respiration is when a..." Gavin stopped, realizing he was not sure of the difference.

Since no one else was jumping in to help Gavin, Lauren, having recently discussed some of this in her science class explained, "Respiration is when the cell produces energy from sugars. Remember, this is what happens in the mitochondria. Breathing is the physical intake of oxygen and release of the wastes from respiration. Respiration is a chemical activity because new substances are produced, while breathing is a physical process. It is just the mechanical intake and outtake of substances"

Lauren's friends were pretty impressed with her knowledge.

"What wastes are produced from respiration?" Asked Gavin.

Hillary was listening and since she also remembered what she learned in her science class, she replied. "Gavin, the wastes we exhale are carbon dioxide and water vapor. On a cold day, you can see the water vapor condensing into a cloud near our mouth as we exhale."

"So that is the cloud in front of my mouth on a cold day," said Gavin applying his new knowledge.

"Another difference between Amoeba and Daphnia is Amoeba move by forming pseudopods and Daphnia move by beating their antennae which must be moved with muscles.

"How do you know all this stuff? No feet, just antennae to move. Seems weird to me," said Grant.

"Amoeba have a food vacuole to digest food, while Daphnia have a gut," said Lauren, continuing the comparisons.

Hillary decided to summarize. "It looks like Daphnia and Amoeba have many similarities because they are both alive, yet they have differences in how they carry out their life activities."

Just then Hillary's mom walked in and said, it is getting late and tomorrow is school so it is time to head home. All of Hillary's friends live in the same apartment building as she so it was easy to head home.

Multiplying Things

Monday was Lauren's favorite day in school because the science club meets after school. She knew her friends, Gavin, Grant and Hillary would be there too in addition to Molly and Linc. In science class her lab group, which included Molly and Linc were to set up an experiment to see what factors affect the growth of a plant. They chose to study the Kentucky Wonder Bean plant.

"What factors do you want to test?" Asked Linc.

"Mrs. Squires said we should brainstorm first before we set up an experiment so let's first list all factors that we think can impact the growth of the plant," said Lauren.

"Crowding could prevent some plants from getting the needed minerals it needs to grow properly," suggested Molly.

"The amount of water can impact growth," added Lauren.

"The amount of sunlight," said Linc.

"The type of soil we use," stated Molly.

"How often we fertilize said Lauren.

When they could no longer think of any other factors, they decided to test crowding.

Mrs. Squires was walking around and observing what the different lab groups were discussing. "What will be your hypothesis for crowding?" She asked their team.

Lauren being the most outgoing of the group responded, "The crowding will result in some seeds germinating and others dying. Because some seeds are able to germinate faster than others, they may shade out the smaller plants."

"They may also use up the available water faster than the smaller plants," added Molly.

"Excellent hypotheses, said Mrs. Squires. "Now set up your experiment and see what happens!

The team set up three plant boxes. They noted their dependent variable would be the number of plants that grow and produce beans, while the independent variable is the number of seeds they plant in each box.

"Now we must keep everything the same for all of our groups," stated Lauren.

"We should use the same soil, watering times, amount of water, box size and place all in the same amount of sunlight," said Linc.

After all the lab teams in Mrs. Squires' class set up their experiments, she led a discussion about what factors affect the success of an organism reproducing.

Lauren remembered that Daphnia protect their eggs in a pouch so she suggested that to her class.

Mrs. Squires responded to Lauren saying that was an excellent example and then told the class how a female seahorse lays her eggs in a pouch that is carried by a male seahorse.

"Why do you think it is helpful for the eggs to be carried in a pouch instead of left to the open?" She asked her class.

Molly responded, "Because if the eggs are not protected they can be more easily eaten by predators.

Shanelle, a student in Lauren's class said that some living things take care of their young and build nests.

Linc said, "some plants such as oak trees produce a lot of acorns. That helps to ensure that some will grow into a tree because so many will be used for food by other living things."

Mrs. Squires asked the class how animals attract one another for mating.

Lauren said, "Well in the case of birds, the male is usually more colorful than the female and that attracts the female for mating.

"I heard of fish that puff up to attract a mate," said Molly.

"Some do a dance," added Shanelle.

"Even their smell can attract a mate," added Linc.

"I once saw a nature movie on TV. It showed how elephants keep their babies in the middle of a herd when traveling. The show said it was a way in which the elephants work together to protect their young from predators," explained Lauren.

"Excellent examples!" Mrs. Squires said complimenting her class. "All living things have strategies to make sure their offspring survive. Today, you all set up an experiment with plants. What factors do you think will affect how well a plant grows and develops?

Lauren had her hand up immediately. Being a wise teacher, Mrs. Squires waited a bit to give other students a chance to respond. She decided to call on Molly first.

"The amount of water and sunlight will affect how well a plant grows," replied Molly.

After complimenting Molly, Mrs. Squires then called on Lauren.

Lauren said, "The genetics of a plant will also affect how well it grows. Some may be able to grow faster.

Shanelle raised her hand and when called upon said, "The color and odor of flower petals attracts insects so pollen can be transferred to other flowers and seeds can then be produced inside the flower. So, can't you say both a plant's heredity and environment affect its development. If there were no insects, pollination would not take place."

"Absolutely," replied Mrs. Squires, "Scientist sometimes refer to this as 'nature vs. nurture. Nature referring to an organism's heredity and nurture would be the environment in which an organism lives."

Shanelle raised her hand again and stated, "So my friends Sasha and Michelle are identical twins, but Sasha is taller, so could I say the difference is due to nurture."

Mrs. Squires paused a minute because she had to think through this question. She replied, "Yes, no doubt nurture played a role, but even though they are identical twins, there could be some genetic differences we do not know about. Scientists are always learning new things about how our genes operate."

Just then the bell rang and the students proceeded to their next class.

Lauren could not wait for the bell to ring because it was her last class of the day and science club would be meeting next. Though she enjoyed her social studies class, she was happy when the bell rang. Lauren met up with her friends, Grant, Hillary and Gavin in science club. The advisor to the club, Mr. Storey told the students that they were going to do conditioning experiments with mice with the mazes they had built. The science clubs spent three meeting researching and building different types of mazes.

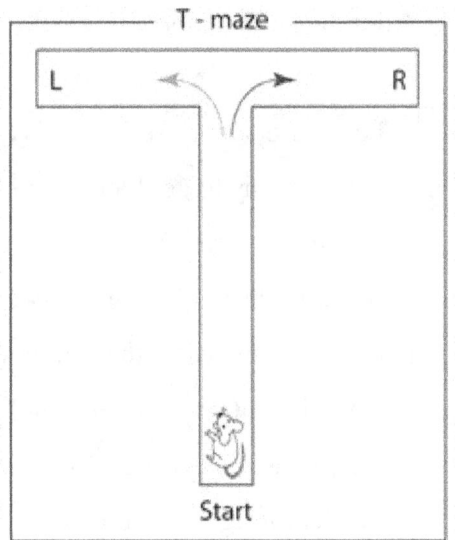

Lauren and her friends built a T-maze. They were going to train a mouse to run the maze and always make a right turn to get food. They were taught how to handle mice and their advisor did not feed the mice that day until the club met. Hungry mice would be more active and more likely to explore a maze. Gavin got their mouse whom they named Snuggles. They placed Snuggles at the beginning of the T-maze. As expected,

Snuggles explored the maze. His brain was stimulated by the touch of the maze. Eventually, Snuggles had to make a decision to turn right or left. The friends had placed food in the right chamber. Snuggles turned left first, paused and then ran back to the beginning of the maze.

"Snuggles, the food is on the right," coaxed Gavin as if the mouse could understand him.

Snuggles went down the maze and turned left again.

"Oh boy, this is going to take a while," said Grant.

However, Snuggles then went to the right chamber and found the food, which he quickly ate.

Hillary took the mouse out of the maze and held him for a minute and then placed him back in the maze for a second trial. Grant kept time for how long each trial would take and how many wrong turns the mouse made each trial.

By the 8th trial, the Snuggles went to the right chamber without any errors. They did the experiment three more times and Snuggles made no errors.

"Snuggles, you are so smart," said Lauren.

"We are excellent mouse trainers," said Gavin.

Mr. Storey had all the club teams report the results of their conditioning experiments.

After all of the reports, he asked the club what they learned.

Hillary said, "The mouse formed an association between turning right and getting food."

"The mouse's brain sent signals to nerve cells that send signals to muscles for the mouse to turn the correct way to get food," said Gavin.

All of the club members agreed that the mouse had learned their maze through this conditioning experiment, though some teams, more successful than others.

Hillary, remembering the fun time she had with her friends looking at pond water said to everyone. "All living things can respond to stimuli in the environment. A plant moves toward the sun, a predator chases its prey and now we taught a mouse to run a maze to get its food."

"Excellent insight," responded Mr. Storey to Hillary's comment. "I think all of your brains learned a few things today too and I look forward to meeting with all of you next week when we study the behavior of a land snail."

The club members cleaned up. Lauren, Gavin, Grant and Hillary walked to the bus stop together since they lived in the same apartment building.

"What do you want to do next weekend," asked Gavin of his friends.

Staying Alive, **Tales of Science** by Joan S. Wagner

"Let's dissect a fish," said Grant.

"I have a better idea," said Hillary, "why don't we put together a plan to build a pond at our school. Our science club would sponsor it. We can then bring the plan to Mr. Storey.

Hillary's friends thought that was a much better idea than dissecting a fish so now they had a plan for their end of the week get together at Lauren's place this time.

Discussion Questions
1. Why can a car not be considered a living thing?
2. Describe three characteristics of living things.
3. How is a unicellular organism and multicellular organisms alike and different?
4. What is the purpose of an amoeba's pseudopods?
5. What are some ways in which living things respond to their environment?
6. How can overcrowding impact living things?
7. What is the difference between a dependent and independent variable?
8. In the overcrowding plant experiment, what are some factors that should be held the same for the different experimental groups the children set-up?
9. All living things can respond to the environment. Why do plants move toward the sun?
10. Why did the mouse learn to turn the right way to get food?
11. What is meant by nature vs nurture?
12. What are some factors that can affect the growth of plants?

Staying Alive Science Terms

1. *Amoeba*: A single-celled organism found in pond water
2. *Binary fission*: at type of reproduction called asexual since no sex is involved. A single-celled organism reproduces by dividing in half.
3. *Breathing*: The Ability to take in oxygen and release carbon dioxide and water vapor.
4. *Cell*: The smallest unit if a living organism
5. *Cell membrane*: The outer coating of a cell that regulates what goes into and out of the cell.
6. *Cell wall*: The outer coating of a plant cell, often composed of cellulose and gives a plant cell its shape.
7. *Chlorophyll*: The green pigment in a photosynthetic cell that uses the energy from the sun to make food.
8. *Chloroplast*: The structure inside of a photosynthetic cell that contains the chlorophyll.
9. *Cytoplasm*: All the jelly-like matter in a cell that holds its organelles
10. *Contractile vacuole*: A structure inside a cell that acts like a kidney regulating ridding the cell of liquid wastes
11. *Daphnia*: A water flea
12. *Digestion*: The process by which food is broken down chemically to be used by a cell.
13. *Egestion*: The process of ridding of undigested wastes.
14. *Egg*: A cell used for reproduction
15. *Excretion*: Ridding of digested wastes.
16. *Eyespot*: A pigment spot in some cells that is sensitive to light.
17. *Flagella*: A organelle used for motion
18. *Food vacuole*: An organelle that functions like a stomach
19. *Gills*: A structure used for breathing.
20. *Green algae*: a type of single or multicellular organism that can make its own food.
21. *Gut*: A stomach-like structure.
22. *Mitochondria*: Known as the power house of the cell and is where the cell makes energy for its life activities.
23. *Multicellular organism*: A n organism compose of more than one cell.
24. *Nucleus*: The control center of a cell.

25. *Organelle*: a structure in a cell that carries out a life activity.
26. *Photosynthesis*: The life activity by which an organism uses energy from the sun to make food.
27. *Pyrenoid:* Found in chloroplasts of mostly algae and helps with its photosynthesis.
28. *Pseudopod*: a false food used for motion by cells and to engulf and eat other organisms.
29. *Reproduction*: The life activity of making more of itself.
30. *Respiration***:** The life activity of producing energy for cellular activity.
31. *Stimulus*: Something in the environment that causes a response from an organism.
32. *Unicellular organism*: A single celled organisms such as the Amoeba.

Standards included in this story

MS-LS1   From Molecules to Organisms: Structures and Processes

Disciplinary Core Ideas

LS1.A: Structure and Function
1. All living things are made up of cells, which is the smallest unit that can be said to be alive. An organism may consist of one single cell (unicellular) or many different numbers and types of cells (multicellular).
2. Within cells, special structures are responsible for particular functions, and the cell membrane forms the boundary that controls what enters and leaves the cell.
3. In multicellular organisms, the body is a system of multiple interacting subsystems. These subsystems are groups of cells that work together to form tissues and organs that are specialized for particular body functions.

Growth and Development of Organisms
1. Animals engage in characteristic behaviors that increase the odds of reproduction.
2. Plants reproduce in a variety of ways, sometimes depending on animal behavior and specialized features for reproduction.
3. Genetic factors as well as local conditions affect the growth of the adult plant.

Organization for Matter and Energy Flow in Organisms
1. Plants, algae (including phytoplankton), and many microorganisms use the energy from light to make sugars (food) from carbon dioxide from the atmosphere and water through the process of photosynthesis, which also releases oxygen. These sugars can be used immediately or stored for growth or later use.
2. Within individual organisms, food moves through a series of chemical reactions in which it is broken down and rearranged to form new molecules, to support growth, or to release energy.

Information Processing
1. Each sense receptor responds to different inputs (electromagnetic, mechanical, chemical), transmitting them as signals that travel along nerve cells to the brain. The signals are then processed in the brain, resulting in immediate behaviors or memories.

Energy in Chemical Processes and Everyday Life

1. The chemical reaction by which plants produce complex food molecules (sugars) requires an energy input (i.e., from sunlight) to occur. In this reaction, carbon dioxide and water combine to form carbon-based organic molecules and release oxygen.
2. Cellular respiration in plants and animals involve chemical reactions with oxygen that release stored energy. In these processes, complex molecules containing carbon react with oxygen to produce carbon dioxide and other materials.

Science and Engineering Practices
1. Designing and using models
2. Planning and carrying out investigations
3. Constructing explanations and designing solutions
4. Obtaining, evaluating and communicating information

Cross Cutting
1. Cause and effect

We are Alike, Yet Different

Living things are all around you. Some are so tiny that you need a microscope to see them, while others can be easily seen. Some living things look alike and some look different.

This was the first page of a book that Eliza and Oliver's mother gave them. As fraternal twins, they became very curious about what makes them alike and what makes them different.

Cats and Dogs

Since they had a dog and a cat, Eliza and Oliver decided to compare their pets to see how they were alike and different.

"Some people may think cats and dogs are different from one another, but, actually, they have many things in common," Oliver told Eliza.

Though Eliza was reading a Harry Potter book, Oliver continued anyway. "Dogs and cats are alike because they are both carnivores, give birth to live young, produce milk, and care for their young."

Eliza looked up from her book.

"Ahh, you are listening."

"Well, I was enjoying my book until you began your cats and dog talk."

"I thought I made some good points."

"You did, but there are many more similarities you did not mention."

"Like what?"

"Well, you could add that their bodies are covered with hair or fur, both have an excellent sense of smell and can hear a wide range of sounds. They also move on four feet.

"So, put your book down and let's come up with differences between cats and dogs."

"Remember when our cat, Xavier, scratched my arm and made it bleed?"

"And made you cry."

"Well I was only eight when I got scratched and it did hurt," replied Eliza.

"Though cats and dogs both have claws, cats have claws that go in and out."

"That is called retractable claws."

"Right, so there is one difference" noted Oliver.

"How else are they different?"

"Well there must be differences in their DNA, but we can't see those differences without a DNA fingerprint."

"True, but it would be cool if we could."

Thinking of more differences, Oliver said, "Dogs and cats communicate differently."

"Right, I haven't heard a cat barking recently."

"Very funny!"

"So, what do you think makes us alike?" Eliza asked her brother. "I can definitely see our differences. We do not even have the same color hair and I am taller than you, and smarter…oh, just kidding!"

"Well, we walk upright, have ten toes and fingers. We both can grasp things because of the placement of our thumbs. Dogs and cats definitely do not have that trait. Without an opposable thumb, they cannot grasp things like we can."

"It would be funny if they could. They could open their own dog and cat food."

"That is a funny thought, Eliza. Now how are we alike? We are both human beings or *Homo sapiens,* our scientific name."

"Better known as the wise man. Though I think it should be changed to wise person."

"Actually, the word *Homo* is from Latin and means human being," said Oliver.

"Well, that sounds a lot better to me," replied Eliza.

"I agree with you, but let's not get distracted over the issue of boys vs. girls."

The book that was given to Oliver and Eliza had a lot of interesting information about life and how living things adapt to their environments.

Now You See Me, Now You Don't

"Here is a puzzle for you," Oliver started. "What looks like a twig, but has legs and can move?"

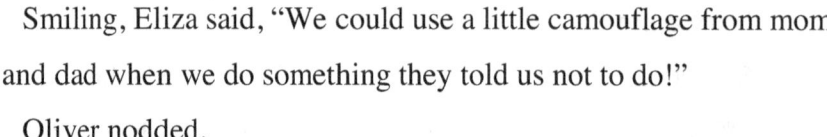
Walking Stick insect

"You must mean a walking stick. It's an insect and even though it looks like a twig, it's an animal not a plant," replied Eliza. "Why do you think it has this trait?"

"That's easy, looking like a twig provides the insect protection from becoming a tasty meal for another animal that likes to munch on insects. For a similar reason, people fighting wars wear camouflage clothing to protect them from being seen by their enemies."

Smiling, Eliza said, "We could use a little camouflage from mom and dad when we do something they told us not to do!"

Oliver nodded.

Can you find the moth?

Later that afternoon, Eliza and Oliver decided to play a Camouflage Game they found on the Internet. In order to play the game, they needed a large piece of wrapping paper with a design, white paper and black paper. They taped a four-foot by four-foot piece of wrapping paper to the floor. Using the template from the Internet, they cut out 10 butterflies that were black and 10 that were white. They also used the wrapping paper (not the paper taped to the floor) to cut out 10 newspaper butterflies.

These were the directions: (See next page).

Make 10 copies of this butterfly in black, white and wrapping paper print.

Camouflage Game Rules

Directions

1. Find a clock with a second hand or a timer. Have a person keep time or set the timer for 15 seconds.
2. Place all of the butterflies in a small bag.
3. Close and then shake the bag so that all of the butterflies get mixed together.
4. Dump all of the butterflies onto the large piece of wrapping paper. Make sure they are spread out and are not on top of one another.
5. Have one person keep time. The other person is given 15 seconds to pick up as many butterflies as she or he can. The person will pretend to be a bird that likes to eat butterflies.
6. You can only pick up one butterfly at a time and it must be placed in the bag before you can pick up another one.

7. How many black, white and wrapping butterflies did you pick up? Make a data table to organize the number of each color butterfly that you pick up.

Eliza and Oliver made their data tables before starting to play the game, forming columns for black butterflies, white butterflies, and wrapping paper butterflies. They would write the number they collected after each round they played. In the first round, the siblings quickly found out that the wrapping paper butterflies were harder to find because they blended with their environment. Eliza and Oliver liked the game so much that they played it a number of times, always having similar results. They realized repeating the game was a lot like repeating an experiment and that repeating an experiment helped to ensure their results were more valid. If Eliza and Oliver formulated a hypothesis before they played the game it would have been something like this: 'Butterflies that blend with their environment are harder to find, so fewer will be captured and eaten'. Of course, they understood that some hypotheses might not be supported, but they were pretty sure about this based on observations they made watching living things that have a camouflage trait.

Eliza and Oliver made sure to place all of the butterflies back in the bag before each game replay. Here is an example of a data table that can be used for this game.

Name of player	Number of black butterflies	Number of white butterflies	Number of wrapping paper butterflies

It's Good to be Adaptive

Polar Bears and Brown Bears

Illustration by Arthur Rackham 1918

Oliver and Eliza were putting away the camouflage game when their little brother, Griffen, came running over to them carrying his favorite book, *The Three Little Bears*. He wanted them to read the book to him.

Reading each other's mind, the twins agreed to read the book, but with a twist. They asked their brother if he knew how the three brown bears in the story were different than polar bears. Griffen looked at his siblings, puzzled, but replied that brown bears live in the forest and polar bears live where there is ice. Oliver added that the location was the Arctic and that brown bears hibernate or sleep during the winter months when there is not a lot of food available.

Eliza said to her little brother, "Both bears are similar to each other, but a polar bear's fur is composed of hollow, transparent hairs. A brown bear's hair is not. The hollow hairs reflect sunlight, which makes them appear white."

Alaska Department of Fish and Game

Griffen looked at his sister like she was from another planet. However, this did not distract Eliza and she continued as if she was a teacher providing a lecture. "Though the white appearance gives the polar bear some camouflage on ice, polar bears have few if any natural enemies so they really do not need to hide."

Griffen looked at his older brother hoping that he would read the story, but instead, Oliver continued Eliza's train of thought. He asked, "Griffen, did you know that the hollow, transparent hair is a special adaptation the animal has to help it stay warm?"

Griffen shook his head wondering when his older brother and sister would actually get to the story.

We are Alike, Yet Different, <u>Tales of Science</u> by Joan S. Wagner

Oliver continued as if Griffen needed to better understand adaptations. He next said, "An adaptation is any trait a living thing has that helps it to survive. It is very cold in the Arctic where polar bears live. Though a polar bear's fur appears white, its skin is black. The polar bear's hollow hairs are filled with air that is warmed by the sun and then absorbed by its black skin. Black is an excellent absorber of heat. That is why you should not wear black in the summertime!"

The twins looked at each other knowing that their poor brother had no idea what they were talking about. Oliver and Eliza were having fun putting a little science into a fairy tale so they continued regardless of the strange faces their younger brother was making.

Eliza decided to continue with the bear comparison. "In addition to their thick fur coat, polar bears have many layers of fat that protect them from the cold climate, helping them keep warm. Even the bear's feet have little bumps on them that act like suction cups and helps prevent them from slipping on the ice."

Though Griffen just wanted to read the story, he was finding what his siblings were saying to be interesting so he asked them, "How are polar bears and brown bears alike?"

Eliza answered, "They look alike. Polar bears and brown bears both have heavy bodies that support a large hairy head. They have thick fur and small eyes with poor eyesight. Their small rounded ears stick up, but their hearing is only fair."

"What do they eat," asked Griffen actually showing interest in what his brother and sister were saying.

"Polar bears eat mostly seals while brown bears eat both plants and smaller animals," answered Oliver.

"Polar bears and brown bears are alike, yet different," concluded Eliza.

After this, the siblings read Griffen his favorite story and then their mom prepared him for bed.

Finches and Charles Darwin

Eliza and Oliver continued their discussion on how living things were alike and different.

They thought a book they had recently received could help them better understand why there is so much diversity around the world. Eliza read first:

"Many people enjoy the hobby of bird watching! There are so many different varieties of birds. Some birds, like chickadees, eat seeds while other birds, like robins, eat worms. Ducks are excellent swimmers. They have webbed feet, which help them swim in the water. Even dogs that hunt water birds or pull sleds have webbed feet to help them move over the snow or swim in the water."

"I think that it is cool that dogs can have web feet," Oliver commented.

"Our Siberian Husky, Max, has web feet. It helps him move through the snow," said Eliza.

Next it was Oliver's turn to read from the book:

"A very long time ago, there was a naturalist, someone who studies nature, named Charles Darwin. He lived from 1809-1882. He noticed there were a variety of finches, a type of bird that lived on a group of islands off the northwest coast of South America called the Galapagos Islands. In fact, he identified 13 different finches on the islands. He observed that their beaks were not alike. Some beaks were better suited for eating insects while other beaks were better at pecking at thick seeds in order to open them up, for example.

"Webbed feet, grasping hands, diet, camouflage: the list seems endless in the way living things adapt to their environment," said Eliza. "My turn to read now!"

"Darwin figured that if all the finches on the islands could only eat insects, there might not have been enough for all of the birds. Some finches would go hungry and die. Since the finches had beaks that were suited for different types of food, there was plenty of food for all of the finches. The finches did not have to fight with one another for food because they did not have to compete for the same food. The finches were alike yet different." Eating different types of food helped them survive living in the same region. Darwin would say having different beaks suited for different meals was a way they adapted to living on the Galapagos Isles together.

Their book included an adaptation game to play. They had enjoyed the camouflage game so they read over the directions of the 'Adaptation Game' and played it

Adaptation Game Rules:

"Oven Mitt Finches vs Naked Mitt Finches."

1. Obtain 50 pennies. Each penny represents a nut. Obtain one oven mitt. The mitt will represent the shape of the bird's beak.
2. The hand with the oven mitt is called an "Oven Mitt finch." A hand without a glove is called a "Naked Mitt finch."
3. One person will be an "Oven Mitt Finch" and the other person will be a "Naked Mitt Finch."
4. Place an open box on a floor or table.
5. Spread out the pennies on the floor or table.
6. Make a data table to organize the number of pennies picked up by each bird in 15 seconds.
7. How many pennies can the "Naked Mitt Finch" pick up in 15 seconds? After each penny is picked up, it must be placed in the box.
8. Repeat this with the "Oven Mitt Finch."
9. Which finch is better adapted to eating "nuts?" Explain.

_____
_____
_____

Eliza and Oliver played the game a number of times. They always had the same results.

"The poor Oven Mitt Finch would starve if it had to eat penny nuts. I bet it would be a lot easier for the Oven Mitt Finch to eat cherries," hypothesized Eliza.

"We can use marbles for cherries and run the experiment again," suggested Eliza.

"Not now, maybe later," said Oliver. "But I still think the Naked Mitt Finch would do better than the Oven Mitt Finch. In fact, I bet that the Oven Mitt Finch would eventually die out because it could not compete as well with the Naked Mitt Finch for food," continued Oliver.

"I think the Oven Mitt Finch is cute, but I suspect you are right because that would have been the thinking of Darwin, and Darwin was definitely an expert on how living things survive and evolve or change" said Eliza.

It's in Your Genes

Eliza continued her thoughts about why there was so much diversity in life. She knew that people are alike and different. "After all," she said aloud, "some people are short, while others are tall. Some people have straight hair, and others have curly hair. Some people have light skin, and others have dark skin."

"Some people like to ski and others like to snowboard. Some people like hip hop and others classical music," teased Oliver, knowing those were not really traits.

Then he got more serious. "Look at all the variety of dogs," added Oliver. "Siberian huskies look very different than cocker spaniels. And, just like people, there are big dogs and little dogs."

"Just think about all the different types of dogs that we know about," said Eliza.

"And think about this," said Oliver. "When a beagle mates with a German Shepard, the puppies look like both parents, but are not identical to either parent."

"It sure would be funny if the head was a German shepherd, and the rear was the poodle," said Eliza, "but that does not happen."

"Right, all living things become a mixture of both parent's traits though some traits may dominate others."

"What's a trait?" asked Griffen.

"We thought you went to sleep," said Oliver and Eliza in unison.

"I wanted to give both of you a hug goodnight," Griffen told them.

Eliza and Oliver gave their little brother a hug goodnight. Then Oliver answered Griffen's question. "A trait is something that is inherited. Your blue eyes are an example of a trait. It is passed on from parents to their children. Another name for inherited traits is genes.

"Like what we wear," said Griffen.

Eliza replied, "No, these are not what you wear! Those are j-e-a-n-s. The genes you inherit are spelled 'g-e-n-e-s."

"Oh," said Griffen.

"Time for bed, Griffen," called their mom. His older siblings watched as he went to bed rubbing his tired eyes.

Oliver said to Eliza, "The fact that you have blue eyes and I have brown eyes is an example of how some inherited traits can be masked or hidden. Mom has blue eyes and Dad had brown eyes. You have blue eyes because you received one gene for blue eyes from mom and another one from dad. Though dad has brown eyes, his brown-eyed gene masked the presence of the blue-eyed gene. Mom has blue eyes because both of her genes are for blue eyes. I am like dad."

You sure remember a lot from Mrs. O'Leary's 7th grade science class. I forgot a lot of that," noted Eliza. She decided to challenge her brother.

"What about green eyes?' asked Eliza. That question stumped Oliver.

Their mom was listening to them talk as she returned after putting Griffen to sleep. She explained that they were basically correct, but that eye color inheritance is a little more complicated since there are more than two types of genes involved. But she did say that they had the general idea as how their eye color was inherited. Green eyes depend on some other genes that control eye color.

Having a mother who is a science teacher is always handy!

How You Look is Like a Game of Chance

Eliza and Oliver were glad they were not identical twins. They did not mind sharing the same birthday, but that was enough. However, they did have identical twin cousins, Farrah and Elle. Farrah and Elle usually had fun pretending to be one another. However, people who knew them could tell them apart. For example, Farrah was a little taller than Elle and had a small birthmark near her chin.

Oliver and Eliza learned from their cousins that identical twins could be mirror images of one another. Farrah is right handed and Elle is left-handed. They did know that identical twins could have many more similarities than they do as fraternal twins.

After listening to her children's interest in heredity, Oliver and Eliza's mom asked them if they would like to play a game that will help them understand heredity better. "It is sort of like a game of chance," she said.

Oliver and Eliza said yes, excited for another game.

As an introduction to the game, their mom said, "Suppose a child named Hunter has blond, wavy hair and brown eyes. His brother, Dylan, has brown straight hair and green eyes. Their mom has green eyes and wavy brown hair and their dad has brown eyes and straight blondish hair.

"They look like they each got some of their traits from both parents," said Eliza.

"Excellent," their mom responded.

She continued, "Children are a mixed-up blend of their parents. How children look can be thought of as a game of chance. Half of Hunter and Dylan's genes came from their dad and the other half came from their mother. Their traits are determined like flipping a coin; it occurs by chance. Even if a child is a boy or girl, it is determined by chance!"

"You mean the chance of my being a boy was the same as being a girl?" asked Oliver.

"Sort of," said his mom. "Let's play the game, "Coin Children," and you will understand how it works better," she continued. "You will determine the traits a girl named Chessie can inherit from her parents."

She took out six coins and then said, "You have to do a little pretending. Pretend that each side of the coins has one variation of a trait Chessie can inherit from one of her parents."

"So, there are two variations for each trait in this game," stated Eliza.

"You are catching on quickly," said her mom.

"Since half of her heredity comes from each parent, you need two coins for each trait. One coin would have the genes, or traits, Chessie can inherit from her mom. The other coin would have the genes Chessie can inherit from her dad."

"How many pennies will you need to determine hair color, hair type, and eye color? asked their mom.

"Three," replied Eliza.

"No," corrected Oliver. "We need six pennies," he explained. "Though there are three traits, each parent contributes at least one gene a trait so there has to be a minimum of six pennies."

"Very good," said their mom. "You have thought this through carefully."

Seeing that Eliza felt bad she did not have the correct answer, her mom pointed out that it was a common mistake made by others.

Their mom explained to them that the game was simplifying inheritance because scientists now know that many traits are controlled by more than one gene and that other factors can affect whether a specific gene is even expressed (shows in their appearance). However, she did explain that hair type could be straight, curly or wavy. She told them that there is a gene for straight hair and a gene for curly hair, but there is no gene for wavy hair. In order for a person to have wavy hair, they have one curly gene and one straight gene. Many genes can mask the presence of another gene, but for hair type if a person gets one of each type of gene a third trait appears. In the case of hair type, it would be wavy hair. She told them scientists call this *incomplete dominance* and they will understand it better after playing the game.

Eliza and Oliver got six pennies. They placed a little masking tape on each side of the penny. Scientists use a capital letter to denote when a gene is dominant (can mask a recessive gene) and a lower-case letter when the gene is recessive (can be masked). They labeled each side of the six pennies in the following way:

Hair color: B = Brown
        b = blond

Hair Type: C = curly
        S = straight

Eye Color: E = brown eyes
        e = blue eyes

"Eliza, get some brown, yellow, and blue markers or crayons," said their mom.

Now that they prepared the coins, their mom took out the game called, Coin Children.

Eliza and Oliver were ready to play the game. They read the directions.

Directions:

1. Prepare the coins
2. Place all of the coins in the paper cup.
3. Shake the cup and pour out the coins onto the floor or a table.
4. Copy the data table shown here onto a clean piece of paper.
5. Complete the data table by filling in the information that is missing.
6. Repeat steps 1-4 at least three more times, to determine the traits of other children in Chessie's family. Chessie has two brothers and one sister.

Coin	Outcome: Chessie
1	Eye color is _____
2	Eye color is _____
3	Hair color is _____
4	Hair color is _____
5	Hair type is _____
6	Hair type is _____

Coin	Outcome: Hunter
1	Eye color is _____
2	Eye color is _____
3	Hair color is _____
4	Hair color is _____
5	Hair type is _____
6	Hair type is _____

Coin	Outcome: Dylan
1	Eye color is _____
2	Eye color is _____
3	Hair color is _____
4	Hair color is _____
5	Hair type is _____
6	Hair type is _____

Coin	Outcome: Kiley
1	Eye color is _____
2	Eye color is _____
3	Hair color is _____
4	Hair color is _____
5	Hair type is _____
6	Hair type is _____

Draw a picture of what the child looks like using the following rules:

EYE COLOR OF CHILD

2 blue eye pennies (ee) = blue eyes

We are Alike, Yet Different, Tales of Science by Joan S. Wagner

1 blue eye penny + 1 brown eye penny (Ee) = brown eyes

2 brown eye pennies (EE) = brown eyes

HAIR COLOR OF CHILD

2 blond hair pennies (bb) = blond hair

1 blond hair penny + 1 brown hair penny (Bb) = brown hair

2 brown hair pennies (BB)= brown hair

TYPE OF HAIR

2 straight hair pennies (SS) = straight hair

1 curly hair penny + 1 straight hair penny (CS) = wavy hair

2 curly hair pennies (CC) = curly hair

Eliza and Oliver colored in the pictures of the four children using the template their mon provided.

"Wow, we sure have a lot of variation with these three traits. No one is identical in Chessie's family," said Eliza.

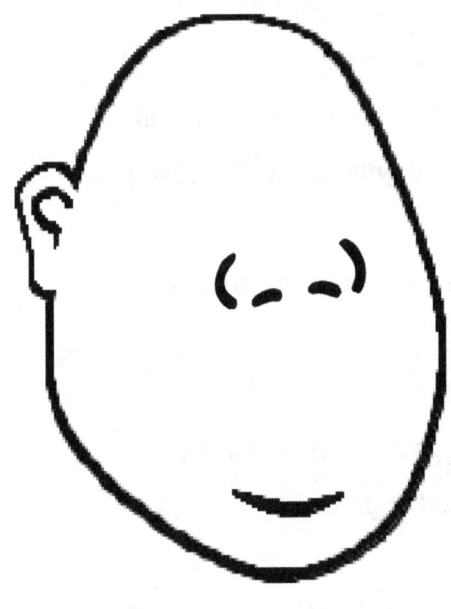

Chessie

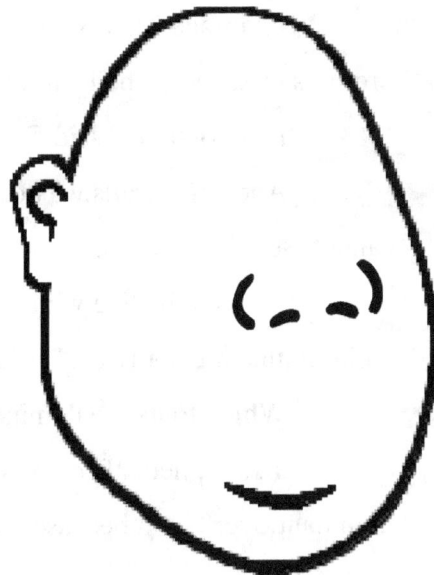

Dylan

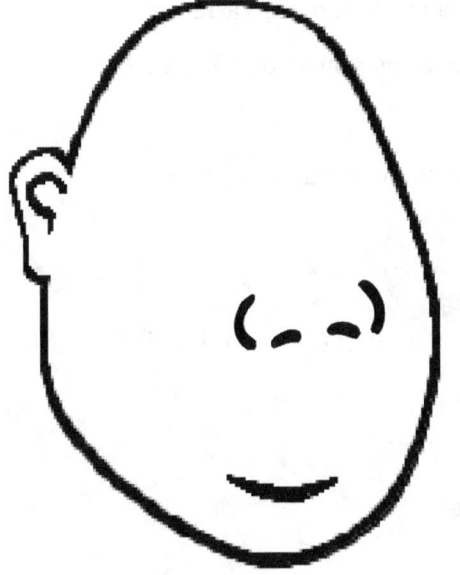

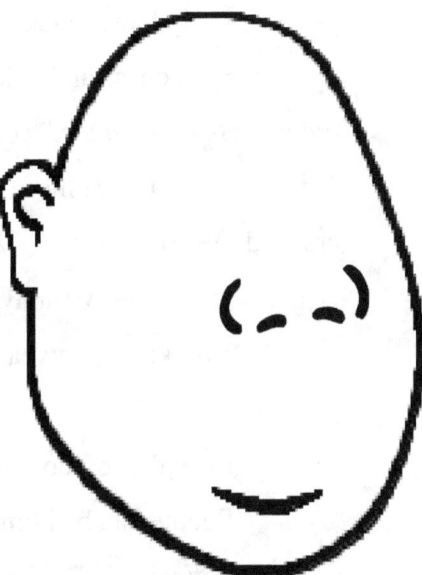

Faces to color to show inherited

Hunter

Kiley

We are Alike, Yet Different, **Tales of Science** by **Joan S. Wagner**

Oliver and Eliza's mother asked them if they thought it was possible for two brothers or sisters to be identical if they were not born identical twins.

"It is possible," said Oliver, "but very unlikely because there are so many traits."

"And some traits we don't even see," said Eliza. "Blood type is inherited, but you cannot 'see' blood type."

"It would be like winning the lottery billions of times in a row to get identical children that are not twins," said Oliver.

"Which traits are dominant?" asked their mom.

Eliza replied, "Brown eyes are dominant over blue eyes and brown hair is dominant over blond because they mask the presence of the other gene."

"Which genes are recessive?" asked their mom

"The genes that are masked, such as blue eyes and blond hair," Eliza answered.

"I think it is neat that wavy hair results from receiving a gene for straight and a gene for curly hair," added Oliver. "Incomplete dominance creates even more variations."

"So, genes can be dominant, recessive or blend together or as you said, it is called *incomplete dominance*," concluded Eliza.

"So, if I have more dominant genes than you do, I must be dominant over you," cracked Oliver.

"Very funny," said his sister.

"You know, you can even predict the chance of getting a certain trait," said their mother.

"How do you do that?" Asked the twins.

"People can be homozygous or heterozygous for a trait."

"Say what?" Replied the twins looking a bit confused.

"Since blue eyes is a recessive trait, the only way a person can have blue eyes is to inherit only the blue-eyed gene. A blue-eyed person is always homogeneous for the trail since a blue-eyed gene was inherited from each parent.

"So, I have a homogeneous trait since I have blue eyes," said Eliza.

"Exactly, and your brother may or may not be homogeneous for brown eyes since your father and I both have brown eyes."

"But how did I get blue eyes?" Asked Eliza.

We are Alike, Yet Different, <u>Tales of Science</u> by Joan S. Wagner

"Your dad and I both must be heterogenous for eye color. Though we both have brown eyes, we each also carry a gene for blue eyes. Since brown eyes is a trait dominant over blue eyes, we both have brown eyes."

"How do we know all of this?" Asked Eliza

"A long time ago, in the early 1800's a monk, called Gregor Mendel did experiments with pea plants. Because of his work, he is known as the father of Genetics or the study of heredity. Let's watch this YouTube: https://www.youtube.com/watch?v=Mehz7tCxjSE

"Now the cool thing about this is that you can actually predict the chance have having a specific trait. AS shown in the video, you can use a Punnett square"

"Can we try one?" Asked Oliver.

"Sure, let's create a Punnett square to demonstrate the chance of a mom and dad heterogeneous for brown eyes having a blue-eyed child.

Their mother completed this Punnett Square for eye color. The top horizontal line and left vertical column shows the possible eye color gene each parent could contribute. She explained that since both parents are heterogeneous for eye color, they both can contribute either a blue-eyed gene or a brown-eyed gene. She then asked her children:

"What is the chance of having a blue-eyed child?"

Eliza was first to answer: "25% because out of the four possibilities only one in four has blue eyes."

"Excellent," said her mom impressed with Eliza's answer and explanation.

	B	b
B	BB	Bb
b	Bb	bb

"What is the chance of having a brown-eyed child, she asked next."

Eliza started to give an answer but she saw her brother wanted to try.

Oliver said, "It looks like three out of four children could have brown eyes so the chance must be 75% and 25% are homogeneous for brown eyes and 50% are heterogeneous.

"You guys are catching on quickly."

The Alphabet Soup of Life

It's All in Your DNA

"Do you know what makes up genes?" asked Oliver.

"Yes, DNA," answered Eliza. "They talk a lot about DNA on crime TV shows. They use a DNA *fingerprint or profile*, to help solve a crime, or to find a missing person, she continued."

"You want to hear something funny?" asked Oliver. "If you spit out your gum and stick it on a desk, a crime investigator could find out whose gum it was because some of your DNA would be in your saliva!"

Eliza laughed. "That would be funny if a teacher actually did that."

"I guess I can use that information to see which one of you leaves crumbs on the kitchen counter every morning," joked their mom.

"But we would have to leave some saliva with our crumbs," noted Eliza.

"Very astute," said their mother.

"What does DNA look like," asked Oliver.

The two of them went back on the Internet and Oliver read the following from a web page:

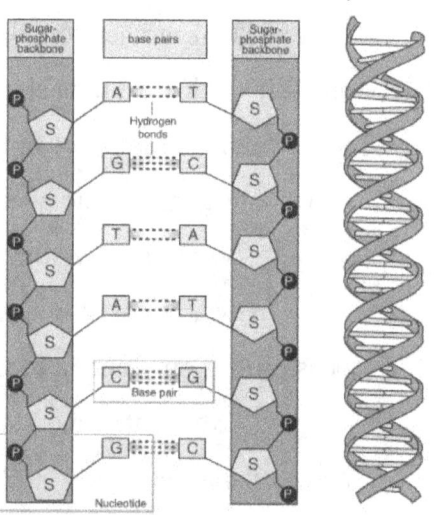

DNA – structure and composition

"Your genes, and the genes of all living things are made up of very large molecules called DNA. Under a very special microscope, a DNA molecule looks like a twisted ladder or double helix. You can think of DNA as having an alphabet of 4 letters: A, T, G, and C (See science terms for more information). The DNA letters can repeat themselves just like they do in words, and the "words" can be composed of a lot of letters or a relatively few letters. For example, they can read AAACGCGATTAGGC or CCGGTAT. Just like we can write many words with the same alphabet, these letters can create many different traits. That is why there are so many different forms of life and why you do not look exactly like either one of your parents. You can think of DNA as being the alphabet soup of life."

"I like that metaphor for DNA being the *alphabet soup of life*," said Eliza. "Next time my English teacher asks for an example of a metaphor, I will use that one"

We are Alike, Yet Different, **Tales of Science** by Joan S. Wagner

You Have a Lot of DNA

Eliza and Oliver read more about DNA on the web. Eliza read out loud this time. "Your body is composed of millions and millions of cells. Every cell in your body has the same exact DNA. In fact, each cell of a human body contains over 3 billion DNA letters. From these letters, scientists have identified about 25,000-30,000 genes. Here is a bit of trivia for you. If the DNA in all of the cells of your body was spread out into a long ladder, it can reach the Sun and go back to Earth six times! Now that is a lot of DNA.

A DNA Fingerprint

"I wonder what a DNA fingerprint looks like," Eliza thought out loud.

Their mom interjected, "Though it is called a DNA fingerprint, it is not like a fingerprint from your hand. It actually looks like bar code."

"Are DNA fingerprints just used for solving crime scenes?" Eliza asked.

"No, there are many uses," replied their mom. "For example, scientists once debated whether a panda bear is part of the raccoon family and not really a bear. If you ever looked at a panda bear, you have to admit that it does look like a raccoon. Scientists were able to use DNA fingerprints of panda bears and raccoons to settle the debate. Panda bears are indeed more closely related to bears and so are correctly classified as bears. However, scientists did learn that the panda bear's cousin, a red panda bear, is more closely related to raccoons."

"So, you can tell which organisms are related," said Oliver.

"Right," said their mom. "If we had a DNA fingerprint of your father and a DNA fingerprint from you, half of your DNA would match your father."

"And the other half would match you," added Oliver.

"Now you are catching on," their mom said.

Oliver and Eliza's mom found a picture of a DNA fingerprint to show them. She then read to them what the book said about how a DNA fingerprint is formed.

"A DNA fingerprint consists of different sized bands of DNA of. Bands of DNA are formed in the following way:

- Scientists first cut up DNA molecules into smaller pieces. They do this with special molecular scalpels called enzymes. Just like enzymes digest your food, they digest or cut up DNA into smaller pieces.

- Cut up pieces of DNA are placed in the wells of a jelly-like substance called agarose gel.
- The agarose gel is placed in a fluid that carries an electric charge.
- An electric current is passed through the gel.
- A DNA fingerprint forms when the cut-up pieces of DNA move toward the opposite end of the agarose gel.
- Big pieces of DNA move more slowly than smaller pieces. Just like there are patterns of runners on a racecourse, a pattern of DNA bands form on the agarose gel. This pattern is what we call a DNA fingerprint or DNA profile.

The reason the DNA moves through the agarose gel is because DNA has a negative charge. The side of the agarose gel opposite the wells has a positive charge. In physics there is a law called, "the Law of Charges." Simply stated, it means that opposite charges are attracted to one another. The negatively charged DNA is attracted to the positively charged end of the agarose gel."

Oliver and Eliza's mom showed them a picture of three DNA fingerprints. She asked them to decide if the possible mom and possible dad were the actual parents.

Eliza and Oliver looked over the DNA pictures carefully. They already knew that half a person's DNA came from their mom and the other half came from their dad. They saw that half of the bands of DNA matched the mom and half matched the dad.

In unison they both said, "Yes, those are the parents."

"Half of the bands match the potential mom," said Eliza.

"And half match the potential dad," added Oliver.

DNA Fingerprint (Profile) on Agarose Gel

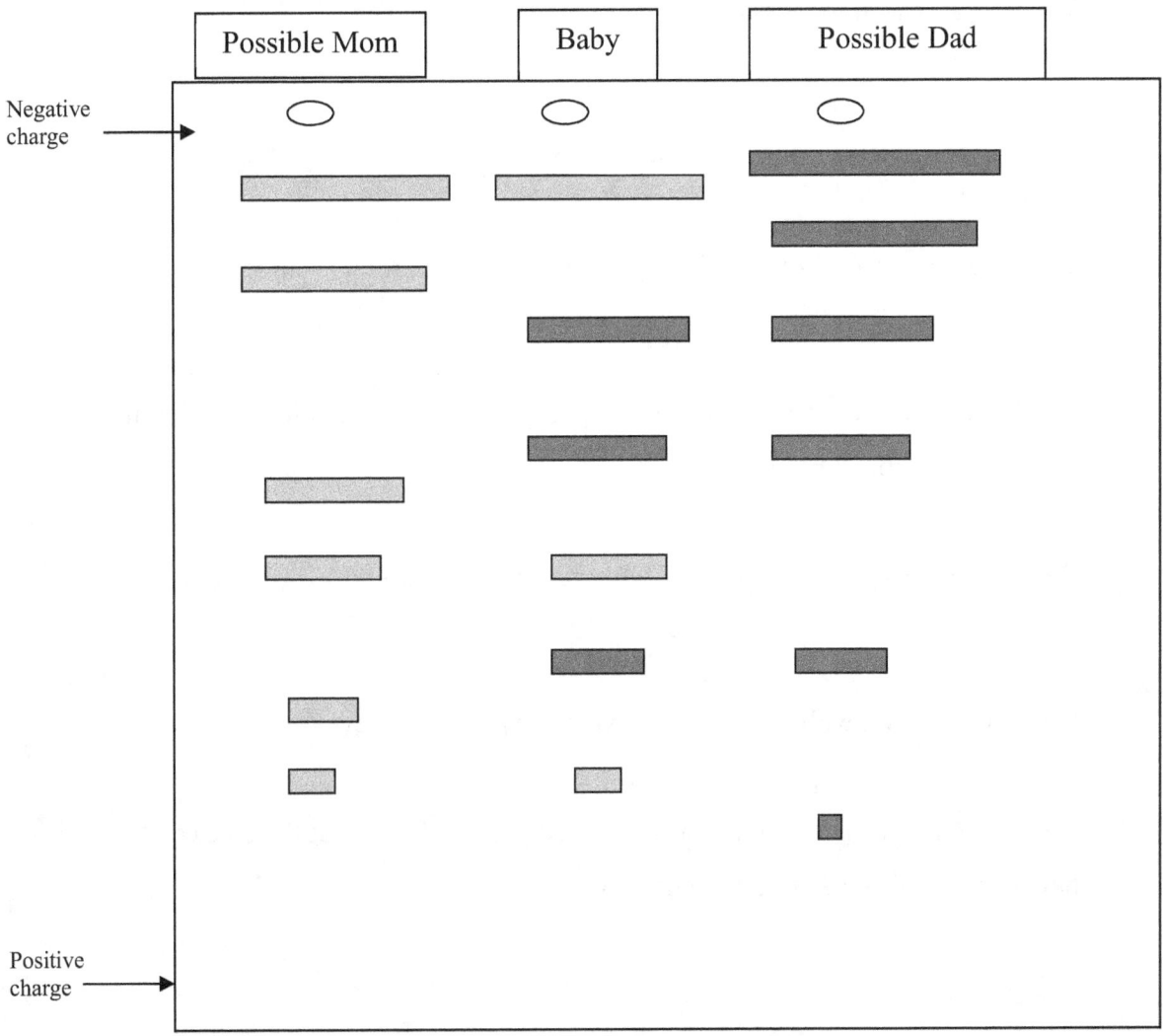

Oliver and Eliza noticed that the baby shares three bands of DNA with its mother (pink bands) and three bands of DNA with its father (blue bands). Identical bands of DNA are the same size and are located the same distance from the wells in the agarose gel.

"I guess it is correct to say that DNA is what makes living things alike and different."

DISCUSSION QUESTIONS

1. What are at least three ways in which all living things are alike?

2. What are at least three ways in which living things differ?

3. What determines the traits living things have?

4. What are some ways in which living things adapt to their environment?

5. What did Darwin observe on the Galapagos Islands that helped him develop his theory of evolution? Explain

6. Describe two ways in which DNA fingerprints can be used?

7. How can a person with brown eyes have a baby with blue eyes?

8. Create a Punnett square for two wavy haired parents. Tell what the chance of having each hair type (straight, curly, wavy).

We are Alike, Yet Different Science Terms

1. *Adaptation:* A trait that has changed over years to help an organism better survive in its environment.
2. *Agarose Gel:* A substance through DNA moves when creating one type of DNA profile that uses electrophoresis.
3. *Camouflage:* The ability of an organism to blend with its environment.
4. *Carnivore:* An organism that eats meat.
5. *DNA:* The substance of which genes are composed. The Letters A,T,G and C correspond to names of the four nitrogen bases found in the nucleic acids that make up DNA, Adenine (A), Thymine (T), Guanine (G) and Cytosine (C).
6. *DNA Fingerprint*: A genetic profile of an organism using their DNA
7. *Fraternal Twins*: Twins that come from different eggs and sperms
8. *Dominant Trait*: A trait that can mask the presence of another trait.
9. *Genes:* Inherited traits.
10. *Gregor Mendel:* An Austrian monk that lived 1822-18844. The Father of Genetics.
11. *Heterogeneous:* Has unlike genes for a specific trait
12. *Homo Sapien*: The scientific name of modern human beings.
13. *Homogeneous*: Has the same set of genes for a trait
14. *Identical Twins*: Twins that come from the same egg and sperm.
15. *Incomplete Dominance*: When two genes for the same trait do not mask one another but forms a new trait.
16. *Inheritance:* The genetic information passed on from parent to offspring.
17. *Law of Charges*: Like charges repel and unlike charges attract.
18. *Masked Trait*: A trait being masked by a dominant trait.
19. *Naturalist:* A person who studies nature such as an environmentalist.
20. *Prehensile Hands*: A hand with opposable thumbs better for grasping objects.
21. *Recessive Gene*: A gene that is not expressed when there is a dominant gene for the same trait.
22. *Retractable Claws*: Claws that can hide inside the animal's paw.
23. *Trait:* A characteristic of an organisms such as eye color.

24. *Transparent*: Anything through which light can pass.

25. *Variations*: The changes in heredity that occur over time.

NGSS Disciplinary Core Ideas

MS-LS1 From Molecules to Organisms: Structures and Processes

LS1.A: Structure and Function

All living things are made up of cells, which is the smallest unit that can be said to be alive. An organism may consist of one single cell (unicellular) or many different numbers and types of cells (multicellular).

Growth and Development of Organisms

Animals engage in characteristic behaviors that increase the odds of reproduction.

MS-LS3 Heredity: Inheritance and Variation of Traits

1. Organisms reproduce, either sexually or asexually, and transfer heir genetic information to their offspring.

2. *Inheritance of Traits*
   - Genes are located in the chromosomes of cells, with each chromosome pair containing two variants of each of many distinct genes. Each distinct gene chiefly controls the production of specific proteins, which in turn affects the traits of the individual. Changes (mutations) to genes can result in changes to proteins, which can affect the structures and functions of the organism and thereby change traits.
   - Variations of inherited traits between parent and offspring arise from genetic differences that result from the subset of chromosomes (and therefore genes) inherited.

3. *Variation of Traits*
   - In sexually reproducing organisms, each parent contributes half of the genes acquired (at random) by the offspring. Individuals have two of each chromosome and hence two alleles of each gene, one acquired from each parent. These versions may be identical or may differ from each other.
   - In addition to variations that arise from sexual reproduction, genetic information can be altered because of mutations. Though rare, mutations may result in

changes to the structure and function of proteins. Some changes are beneficial, others harmful, and some neutral to the organism.

Science and Engineering Practices:
1. Using Models
2. Planning and carrying out investigations
3. Constructing explanations and designing solutions
4. Engaging in argument from evidence

Crosscutting Concepts
1. Cause and Effect
2. Scale, proportion, quantity
3. System and system models
4. Energy and matter
5. Structure and function

Life is A'Changin'

Drew, Noah, Crystal & Caitlin were excited about their field trip to the Science Center because of the dinosaur exhibit.

"I can't wait to see the dinosaurs," said Noah.

"My friend Noe was there and she said they are very life-like. There is even a mama dinosaur with her babies hatching," said Crystal.

"Do you know, there is an animal in Indonesia that looks like a dinosaur. It is called a Komodo dragon," said Caitlin.

"Oh, they are awesome! I have seen them at the zoo," added Noah.

Mrs. Marshall, their 8th grade teacher, led the class into the Science Center.

She said to her students, "Dinosaurs roamed our planet over 65 million years ago, way before people began to populate our planet. In fact, for about 200 million years, they were the major form of life. Now that is a very longtime to be hanging out on our planet!"

"Did humans live then?" Asked Drew.

"No, replied Mrs. Marshall, "People have been roaming Earth for a much shorter time. Scientists are still debating how long modern humans have been around. The best guess is that humans who look like you, your friends and family have been around for 50

Komodo Dragon

Evolution on the Web

Extinction of Dinosaurs:
Go to: Who Killed the Dinosaurs?
http://www.pbs.org/wgbh/evolution/extinction/dinosaurs/index.html
And
Dinosaur Extinction
https://www.enchantedlearning.com/subjects/dinosaurs/extinction/Asteroid.html

thousand years or so. This may seem like a lot of birthdays, but compared to dinosaurs, humans are the "baby creatures" of planet, Earth."

The Dinosaurs Rock!

"What happened to the dinosaurs? Why did they go extinct?" Asked Crystal.

Their teacher saw this question as an opportunity for a lesson so she said, "Generally, living things die out when their environment can no longer provide them with the food and shelter that they need.

The moment of impact 65 million years ago near

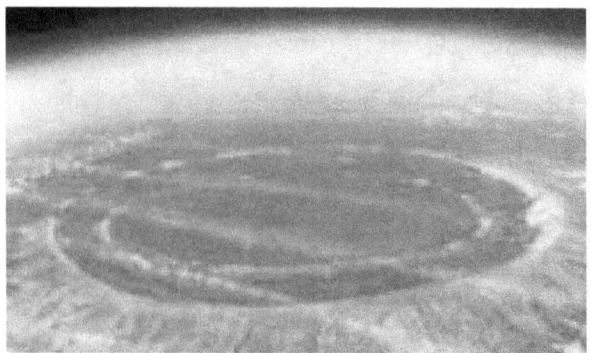

Chicxulub crater, a few days later. Note the inner ring. (NASA artist)

"I heard that an asteroid hit Earth and killed the dinosaurs," said Caitlin.

"Very good," said their teacher, "that is the accepted hypothesis by most scientists.

"Look," said Noah, "It explains what happened here."

"Why don't you read what it says to the class," said Mrs. Marshall since not everyone could see the display with the explanation at the same time.

So, Noah read the following,

"The impact of the asteroid was like the explosion of 100 trillion tons of TNT, a billion times more than the atom bombs that destroyed Hiroshima and Nagasaki."

"Wow, I will have to tell my social studies teacher that," said Caitlin, intrigued by such a huge explosion. Her class had discussed the bombing of Hiroshima.

Noah continued, "It was devastating, causing a gargantuan crater to form in the Gulf of Mexico, located at the tip of the Yucatán Peninsula." Noah pointed to the location on the map that was displayed in the Museum.

Noah used the pronunciation key to read the name of the crater. "Chicxulub (pronounced Chik-shoo-loob), as this impact crater is called measures 120-mile-wide (180 km), 1-mile-deep (1600 m). Based on this information, scientists

Life is A'Changin', <u>Tales of Science</u> by Joan Wagner

were able to figure out the asteroid's speed and size when it smacked into our planet. Its speed was about 20,000 mi/hr (10 km/sec) and its diameter was at least 6-miles (10 Km)."

"Now that is a big cruising rock!" Said Drew.

Noah continued to read the remaining information in the display. "It was a very bad day for the dinosaurs, not to mention other forms of life that lived at that time. Giant fires erupted, incinerating life, lakes evaporated, dust storms blocked sunlight and volcanoes erupted."

"It sounds like a horror movie," said Caitlin, "Like that movie about the world coming to an end."

"It certainly was," responded Mrs. Marshall. Since they had been discussing climate change in class, she thought this was a good opportunity to illustrate how a dramatic change in climate can impact life.

"We have had discussions about climate change," said Mrs. Marshall to her class, "After the asteroid struck Earth, imagine having the climate you once knew changing so suddenly."

"Continue reading," said Mrs. Marshall to Noah.

So, Noah read more from the Science Center display. "Our planet's environment became a very unfriendly place for the dinosaurs and many other forms of life at that time. The fossil record shows that over 70% of all species went extinct after the impact. Recent evidence, using radioactive dating, suggests the dinosaurs went extinct a lot faster than previously thought. They disappeared over a period of 33,000 years as opposed to 300,000."

"That still sounds like a long time to me," said Crystal.

"Yes, that is a long time to us, but our planet is 4.6 billion years old so in geologic time, scientists would call that a short time," explained Mrs. Marshall.

"Oh," said Crystal, "I get it now, it is time relative to the age of our planet."

Noah stopped reading because he had a question of his own. "If scientists are like detectives, what is the evidence that made them so sure an asteroid caused the extinction of dinosaurs?"

Before their teacher could answer, Crystal pointed to another display titled, "Evidence an Asteroid Struck Earth 66 Million Years Ago."

This is what it said:
- The crater left by the impact of the asteroid dates 66 million years, the same time given for the extinction of the dinosaurs.

- The mineral iridium, which is rare on Earth's surface, but plentiful on asteroids and deep inside Earth is 30 times greater in the Cretaceous/Tertiary boundary (K-T), laid down when the dinosaurs went extinct.
- Evidence for once molten rock is plentiful at the K-T boundary. This molten rock also called *impact ejecta* could only have formed from a powerful collision.
- Fractured Crystals also called "shocked quartz," show a distinct pattern characteristic of high impacts or explosions.

"What do you think now? Asked Mrs. Marshall.

Noah thought for a moment and then said, "I see how being a scientist is very similar to being a detective. There needs to be lots of evidence if a hypothesis is to be accepted by scientists. Since there is so much evidence, can the Asteroid Hypothesis now be considered a theory?" he asked his teacher.

Mrs. Marshall replied so the entire class could hear, "Actually, most scientists do refer to it as a theory today because the impact by an asteroid explains many observations.

She continued, "The idea that our planet was struck by a large asteroid was first proposed in 1980 by the scientists Luis Alvarez and his son, Walter Alvarez after they discovered the iridium rock layer at the K-T boundary, a geologic time was worldwide.

"What exactly is the K-T boundary and what is iridium?" Asked Drew, still confused. Mrs. Marshall replied (knowing it was explained in the exhibit), "K is the traditional abbreviation for the Cretaceous period, and T is the abbreviation for the Tertiary period. Therefore, the K-T boundary is the point in between the Cretaceous and Tertiary periods. Geologists have dated this period to about 65.5 million years ago, when the dinosaurs went extinct. Most of you have seen the movie, Jurassic Park. That is the age when dinosaurs thrived. Iridium is an element that is rare on Earth, but abundant in asteroid and meteorites."

Drew thought a minute and then said, "So since the iridium is rare on Earth, but plentiful on space rocks," Alvarez and his son thought the iridium found in the KT layer must have come from something large hitting our planet from out of space."

"You are thinking like a scientist," said Mrs. Marshall. "However, when it was first proposed, many scientists could not agree on whether there was an impact and if it caused mass extinctions. Scientists require a lot of evidence before they will agree on a theory.

Caitlin suddenly remembered something and said, "I heard that dinosaurs evolved into modern day birds."

"Yes, that is what scientists believe to be true based on extensive fossil evidence," said Mrs. Marshall.

Just then the curator of the Museum was walking by and heard her reply. Mr. Hunter introduced himself to the children. He said, "Birds evolved from a group of dinosaurs known as Maniraptoran theropods, generally small meat-eating dinosaurs that include Velociraptor of Jurassic Park fame." He then pointed to a picture in the Science Center.

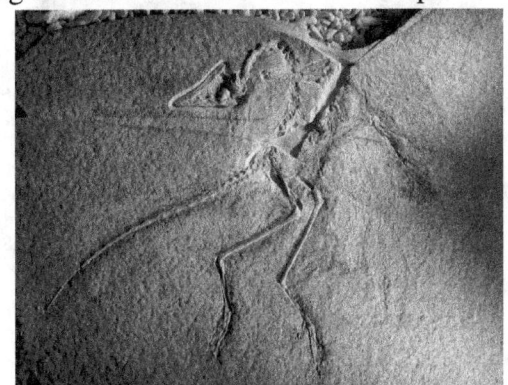

Velociraptor, about the size of a turkey, but very

"I knew those crows in my backyard looked familiar to me," joked Caitlin.

"Still weird to think that dinosaurs evolved into birds," said Noah.

"I just don't get how scientists can come to this conclusion way after the dinosaurs went extinct," said Drew.

"Scientists are like detectives," said the curator, not knowing they were already aware of that. "They noticed that the eggs of these dinosaurs were similar to the eggs of birds. Furthermore, from the position of groups of fossils, they saw similar behavior in how they care for offspring," said Mr. Hunter.

"I thought they were like reptiles," said Crystal. "Reptiles don't take care of their young."

"Actually," said Mr. Hunter, "The fossil record shows these dinosaurs cared for their young. In fact, recently they found some soft tissue with strong evidence they had feathers"

"How does the fossil record tell when things lived?" asked Sharde, another student in Mrs. Marshall's class.

"I am impressed with the inquiring minds of your students," said Mr. Hunter to Mrs. Marshall.

To answer Sharde's question, Mr. Hunter took the children over to a Science Center display that simulated a dig. It showed layers of sedimentary rock and in each layer, fossils could be found.

Mr. Hunter then asked the children which layer had the oldest fossils.

Tell-tale bones shared between the birds and meat-eating dinosaurs. The fused collar-bones (furcula) are in green, and the red ankle bone (astragalus) is attached to the bottom of the orange shin bone (tibia). Notice the four toes, three pointing forwards, one back.

Dinosaurs, the Story of Birds by Dr. Paul Willis

Caitlin reasoned, "Well the layers at the bottom must be oldest because it got covered up by all the other layers so the most recent fossils are the ones found on the top layer."

"Then the fossils found in the same layer, must have lived together at the same time," added Drew.

"You are a swift group," said Mr. Hunter, "But sometimes the rock record is not very tidy. Earthquakes can cause rock to slip, but like a puzzle, scientists can still figure out the layer from which it came."

"Do they match up similar rock just like we match up similar pieces in a puzzle?" Asked Crystal.

"You are on the right track," answered Mr. Hunter. "In addition to doing that, scientists can also date rock by using radioactive minerals found in the rock." Fossils found in rock layers only provide relative dates of fossils. You know which fossils are older or younger by their location in the rock layer. The Colorado River carved out the Grand Canyon exposing rock layers that told the relative history of life on our planet.

Radioactive dating gives more precise dates because radioactive

Grand Canyon National Park from Powell Point on the South Rim, (Credit: Annie Scott, USGS. Public domain.)

minerals decay into non-radioactive minerals at a fixed rate."

So, can you tell exactly when something occurred with radioactive dating?" Asked Crystal.

Mr. Hunter replied, "You cannot give the exact date, but you can tell approximately how long ago a fossil lived in thousands of years. Remember a few thousand years is not very long compared to the age of our planet."

"So relative dating of fossils only tells if a fossil is older or younger while radioactive dating provides a time span in which an organism lived," concluded Crystal.

"Excellent!" Said Mr. Hunter. He went on to compliment the entire class because of their insightfulness and curiosity.

The next exhibit Mrs. Marshall took her class to see was on evolution. The class followed their teacher to the next exhibit.

She said, "Let me introduce all of you to Charles Darwin. He was a naturalist, a type of scientist who developed a theory of how life evolved."

Charles Darwin and the Galapagos Islands

"You all already have learned how dinosaurs evolved into birds. The idea that living things change or evolve into other living things has been around for a very long time."

Mrs. Marshall asked Crystal to read from the exhibit.

This is what she read. "In 1831, Charles Darwin went on a voyage that forever changed our understanding about the origin of living things. He was an English naturalist who traveled on a ship called the HMS Beagle. For five years he cruised through the Galapagos Islands, located 500 miles off the west coast of South America (see map). He made many observations and carefully kept a diary of what he saw. Keeping a dairy is a handy way of remembering things. Of course, in those days, there were no computers or Google searches that could be done. Though Darwin worked in a very low-tech era, he developed a very "high-tech" scientific theory about the origin of life."

Evolution on the Web
Darwin's Diary
Check out this website to learn what he wrote.

Darwin's Diary
http://www.pbs.org/wgbh/evolution/darwin/diary/index.html

The class all looked at the map to see where the Galapagos Islands were located.

Mrs. Marshall pointed out how the Galapagos Islands were off the coast of South America. She added that scientists must be excellent observers and Darwin's excellent observations resulted in this theory of evolution that still holds up today.

Darwin's Finches

"Look at all these finches," called out Noah, "They all have different beaks."

Darwin's Finches, from "Voyage of the Beagle" by John Gould (public domain)

Mrs. Marshall explained to the class that the finches helped Darwin come up with his theory about evolution. She then asked Caitlin to read from the Finches Exhibit.

Caitlin walked up closer to the exhibit and began to read what it said to her class. "Darwin carefully collected a group of birds called finches." Caitlin then pointed to all of the finches. "During the voyage, he did not realize that the birds were all finches because their beak shapes and sizes were so different. For example, some were suited for eating insects while others were better at cracking seeds. After learning that all of the birds were finches, Darwin, being a curious man began to wonder why the finches had different shape beaks on the Galapagos Islands. It was sort of like a puzzle to him and puzzles, though challenging can be very rewarding to solve."

"I love doing puzzles," said Sharde.

Caitlin continued to read from the exhibit. "Since the mainland of South America also had finches, Darwin hypothesized that the finches on the Galapagos Islands must have come from the mainland. He figured that strong winds blew them to the islands. They sort of got lost ending up on the islands instead of the mainland."

"Funny to think that birds can get lost," noted Drew.

"So why did the beaks of the mainland finches change when they became stranded on the Galapagos Islands?" asked Caitlin?

"The answer has to do with the need to eat in order to survive," Answered Mrs. Marshall.

"Caitlin, you know how it is to be hungry. If all of the finches could only eat one type of food such as seeds, there may not be enough seeds to feed all of the finches," said Drew thoughtfully.

"Oh, then they would die from starvation, said Caitlin.

"Right," said Drew.

"Solving this is sort of like solving a mystery," interjected, Noah.

Mr. Hunter, who was listening in on the students said the following. "Darwin already knew that there was always variation in the traits of offspring."

"Just like in my family, not everyone looks the same," said Noah.

"Definitely, not in mine!" Added Sharde.

"So, Darwin must have reasoned some beaks are better suited for getting certain types of food.

"Look," said Caitlin pointing to a picture of a finch, "this finch's beak is best for snagging insects. This sure would help keep their bellies full and ensure that they do not die from hunger."

Noah asked, "How could the finches have so many different beaks?"

Mrs. Marshall explained to the class that Darwin figured these traits are passed on to offspring.

"So, if you are a finch that eats insects, you pass that trait to your babies and if your beak is best for cracking seeds, then you pass on that trait," surmised Sharde.

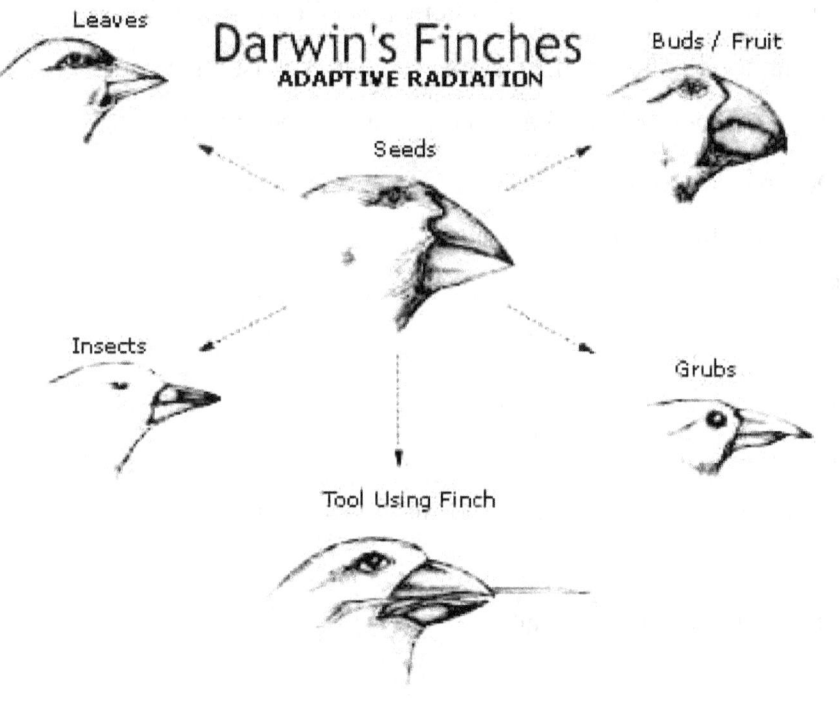

Hudson Audubon Society of Westchester

"So, if you are a finch that eats insects, you pass that trait to your babies and if your beak is best for cracking seeds, then you pass on that trait," surmised Sharde.

"Well if this happens, pretty soon, an entire population of finches would be adapted to eating insects or cracking seeds," concluded Drew.

"Using beak shape for classification, 13 different finches have been identified on the Galapagos Islands," interjected Mr. Hunter.

The class then played the beak game by placing their fingers in a glove shaped like a beak. They would then try to gather food adapted for that shape beak (Directions for this game is in the Chapter, "Alike and Different."

"I can't crack the nut," said Crystal.

"It's because your beak can only eat insects," said Sharde.

"Right," I forgot I was an insect-eating beak.

After playing the game, Noah said, "I like to fish. I think I now understand why some fish have eyes that point downward and others upward. Some are hunting for food on top of the water and some toward the bottom. In this way they don't compete for the same food so there is enough for everyone."

"Very insightful," said Mrs. Marshall. She then told the students, "scientists often refer to these birds as Darwin's Finches because they helped him develop the modern-day scientific *Theory of Evolution by Natural Selection*. In evolution jargon, since all 13 types of finches share the same seed-eating ancestor, scientists call these changes, *adaptive radiation* as shown in the diagram in this display."

Adapting to the Environment

Mr. Hunter walked over to the evolution exhibit to help the students better appreciate Darwin's contribution to science.

He said, "Darwin's adventure through the Galapagos Islands changed the way scientists thought about how life came about or evolved. After all, surviving on our planet is not exactly easy. You have to be tough. You need food and shelter from the cold or heat. Living things must be adapted to their environment or they will die out."

"Become extinct," added Noah.

"Yes, that is the correct jargon," said Mr. Hunter.

"In my household, you have to adapt to mom's cooking or go extinct," joked Drew.

Survival of the Fittest

"I am fitter than you," bragged Sharde to Caitlin. "I can run a lot faster."

"If I wanted to, I could run faster than you, so I would be fitter," retorted Caitlin.

"Girls, please stop arguing," said their teacher.

Mr. Hunter stepped in and said to the girls, "It is true some people can naturally run faster than others but factors such as practice also affect a person's speed. The inheritance of this trait is a bit more complicated in humans."

"What do you think Darwin meant by 'survival of the fittest?" Asked Mr. Hunter of the children.

"Obviously, it cannot mean that living things can intentionally adapt to their environment," answered Drew. It must mean that living things best fit for an environment survive while the others will die out.

"Right," said Mr. Hunter. "For example, the polar bear did not go to the Arctic and think, 'Hmm, it is cold here so I think I will grow a nice warm furry coat or I will become extinct.' According to Darwinian Evolution, the explanation would be that some baby bears already had thick coats that could protect them from the cold. They were able to survive in colder climates while the others died out."

"I would just wear a warmer coat if I was in the Arctic," said Drew.

"Yes, that is true, Drew, because humans can plan and adjust to unfriendly environments. Other animals cannot do that," explained Mr. Hunter.

"What did polar bears evolve from?" asked Noah.

Mr. Hunter replied, "Scientists believe that polar bears actually evolved from brown bears. Just as the Galapagos Islands isolated the finches from the mainland finches, glaciers may have isolated some brown bears from other brown bears. Those bears that were able to withstand the extreme cold and find food survived evolving into the modern-day polar bears."

"Oh, said Sharde, "Then Darwin would say that the environment *selected* the bears that were most fit for the environment sort of like a coach picking the best players for a basketball team so they can survive many games and win the championship."

I have called this principle, by which preserved by the term Natural Selection.

"Excellent," said Mr. Hunter impressed again with how intuitive the children were.

"Isn't Darwin's Theory of Evolution sometimes referred to as the "Theory of Survival of the Fittest?" asked Caitlin.

"Yes," answered Mr. Hunter.

"You have a bright bunch of students here," he said to Mrs. Marshall.

Since he had the attention of the children, Mr. Hunter continued, "Nature has many ways of ensuring that the fittest organisms survive. For example, a herd of zebras is not attacked by a lion because it will only attack an individual zebra. A zebra that cannot keep up with the herd becomes prey for the lion. In this way, you can say that nature has selected speed as a trait needed for zebras to survive. This is what Darwin meant when he discussed *natural selection*."

Variations: Being Different…A Matter of Life or Death

"How do living things obtain new traits that allow them to adapt to the environment?' Asked Drew.

Mr. Hunter answered the question, "Well, take a look at your parents," he said. "If you have brothers and or sisters, take a good look at them too. You must have noticed that they are not exactly like you. Even if you are an identical twin, you probably have some differences.

"I'm identical twin," said Noah, "but I am a little taller than my brother."

"So, you understand what I mean," said Mr. Hunter.

To continue to explain variations Mr. Hunter said, "Suppose you had a sister who was 6"3" tall. Chances are that the basketball coach would be recruiting her to play basketball since height has an advantage for this game. You could say that she is better adapted to basketball than a shorter person. The offspring of living things has *variations*. If a baby zebra was born with a deformed leg, chances are, it would quickly become the meal of a hungry lion once mom stopped looking after it. Deformed legs do not have survival value for zebras. If oak trees produced only a few acorns, the chances that some would make it into a tree would be low. Hungry squirrels would have devoured all of the acorns. Oak trees evolved to produce a lot of acorns, enough for the squirrels and enough to grow into trees."

Bacteria: A Case Study in Variations

Mrs. Marshall then asked her class, "Have any of you after being sick, taken an antibiotic and still remained ill?"

"I have," said Crystal. "The doctor had to try three antibiotics before one worked!"

Mrs. Marshall continued, "For some reason, the antibiotic was not killing all of the bacteria that caused Crystal's infection. Since microbes such as bacteria can breed very rapidly, evolution can speed up because there are so many offspring. The more offspring, the greater chance for variations. Now the bacteria in your body were having a grand time even though Crystal felt awful. They were reproducing like crazy. After taking the prescribed antibiotic, many were killed, but some just happen to have a variation that made them resistant to the antibiotic.

Life is A'Changin', **Tales of Science** by Joan Wagner

Pretty soon, you are feeling pretty rotten again. Now, an entire, new population of antibiotic-resistant bacteria is living it up in your body!"

"So, the bacteria that were resistant to the antibiotic were the fittest and survived and made me continue to be sick," reasoned Crystal.

"Absolutely," replied Mrs. Marshall, pleased with her class discussion.

"So just as the finches developed specialized beaks needed for survival, the bacteria in Crystal developed antibiotic resistance. When she began to take the antibiotic, some of the bacteria were killed, but some bacteria reproduced and formed offspring that were resistant to the antibiotic," reasoned Noah.

"Yes, strange though it may be, evolution took place, on a small scale, right inside Chrystal's body."

"Yuck," said Crystal.

"Now I think I understand why my doctor does not automatically give me antibiotics. If he knows that antibiotics can cause the growth of resistant bacteria, he wants to makes sure I really need the medicine," said Caitlin.

Mrs. Marshall reminded the class that antibiotics are not effective against viruses. "In the past, Doctors overprescribed antibiotics. Now they are more cautious to help cut down on the development of resistant bacteria," she said.

Darwinian Evolution in a Nut Shell

Mr. Hunter asked the class if they could sum up what they have learned about Darwinian evolution.

"Darwin would say that the offspring of living things have *variations*. "They are born with traits that differ from their parents and brothers and sisters, said Drew.

Added Caitlin, "Those with the traits best *adapted* for the environment survive and pass those traits on to their offspring.

"*Natural selection* is the idea that nature selects the fittest to survive," said Noah.

<u>Evolution on the Web</u>
<u>Finch evolution:</u>
http://www.pbs.org/wgbh/evolution/darwin/origin/

"*Survival of the fittest* means that those living things with less suitable traits cannot compete as well and so eventually die out," said Crystal.

Mrs. Marshall was very impressed with her students. She asked Sharde to read a summary about Darwin's finches in the Science Center display.

Sharde read, "When Darwin's finches left the mainland of South America and began to settle on the Galapagos Islands, they became isolated from their mainland relatives, which were mostly seedeaters. Galapagos finches could only breed with other Galapagos finches, most often on the island that they made as their home. Traits that help an organism survive are more likely to be passed on to offspring. This isolation eventually led to the 13 different types (sub species) of finches found on the Galapagos Islands.'

Darwin's Books Caused Controversy

"Look," said Drew. "This display says Darwin's books caused controversy."

"I will try to explain," said Mrs. Marshall to her class. First, she told her class that Darwin published two books. The first one was called 'Origin of Species,' published in 1859. The second book, published in 1871, explained the evolution of humans. It was called 'The Descent of Man.' The first book not only explained the origin of finches, but all life. Since humans are also living things, Darwin felt the need to explain the evolution of humans too so he wrote the second book."

More Evolution on the Web

Check out this website for human evolution:
http://www.pbs.org/wgbh/evolution/humans/riddle/index.html

"But, what was the controversy?" Asked Caitlin.

Mrs. Marshall continued, "When Darwin returned to England, he was very excited about what he had learned even though he knew it would be upsetting to many folks who held a religious view about life on our planet. Being a scientist, he could not ignore his exciting observations about life. However, he wanted to sort everything out carefully before he published his ideas. Therefore, it was not until many years after his voyage that he wrote the books that forever changed our understanding of the origin of life on our planet."

"I am still not getting what the controversy is about," said Caitlin.

Mrs. Marshall replied, "Darwin was a religious man and he knew that his observations, hypotheses and conclusions would be, particularly upsetting to the Church. In fact, this was another reason why he held off having his books published right away. After publication, as he expected, controversy erupted!"

"But what was so upsetting to people?" Asked Drew.

Mrs. Marshall again replied. "In those days, it was the general belief that all life on Earth was created once and does not change. Also, the idea that humans and apes share a common ancestor seemed particularly antireligious to many. In fact, as in Darwin's day, people today still confuse major concepts in evolution. They think that humans evolved from apes. Darwin never said that. He said that based on his knowledge of anatomy and social behavior, humans and apes must have shared a common ancestor. Today, scientists believe that the common ancestor of humans and apes existed about 5-8 million years ago. The ape that is most closely related to us is the chimpanzee."

Darwin Called an Ape

Noah saw a funny cartoon of Darwin looking like an ape in the museum display.

"Read what the display says to the class," said Mrs. Marshall to Noah.

"After the publication of his books, the press had a field day with Darwin and pictures such as this one appeared in many newspapers at that time, making Darwin look like an ape. New ideas in science, particularly those that provide explanations for life that is different than the bible have experienced the most attacks. It is for this reason that the teaching of evolution has been challenged many times in the courts.

"People have sued about the teaching of evolution?" asked Crystal.

Life is A'Changin', **Tales of Science** by Joan Wagner

Evolution Goes to Court

"Yes," said Mrs. Marshall. And she continued to tell the class "Some folks, who do not accept the "Theory of Evolution," have tried to replace it with a nonscientific explanation. *Creationism* and *Intelligent Design* are examples of ways in which some well-meaning people use religion to explain science. The problem is that many people do not understand the *nature of science*. Religion depends on faith. Science depends on testable evidence. They are not meant to contradict one another."

Creationism, Intelligent Design and the Nature of Science

Mr. Hunter thought he would give it a try to further explain the problems schools were having with the teaching of evolution. He said, "Attempts to teach *Creationism* or *Intelligent Design* in the classroom as alternative "theories of evolution" has led to some very nasty battles in the courts. *Creationism* teaches that life has not changed but came about all at once at the same time (5000 years ago). *Intelligent Design* is actually a slightly modified version of *Creationism*. It accepts that life has changed but provides non-scientific reasons as to why it has changed. They are not scientific theories because they cannot be tested in the natural world. Instead, they both are religious explanations for the origin of life on our planet. Because our Constitution separates Church and State (the first amendment right to practice the religion of your choice), it is unconstitutional to teach a religious explanation for the origin of life in a public school. Therefore, when some public-school districts in our country tried to teach religious explanations for the origin of life, many court battles took place between those who wanted it to be taught and those who felt it infringed on our *First Amendment* right."

"Sounds like my social studies teacher should be teaching about these law suits," said Noah. "What happened with the court cases?" Asked Drew.

Mrs. Marshall replied, "The Supreme Court has ruled that creationism is a religion and cannot be taught in public schools. When the Dover, Pennsylvania Board of Education tried to have Intelligent Design taught in their schools, Pennsylvania's Supreme Court also ruled it was unconstitutional."

Mr. Hunter told the students the *nature of science* allows for ideas that were once accepted by scientists to be falsified (proven wrong). This may happen when new observations do not support old explanations. These observations may come from laboratory and field studies of the natural world, computer modeling and other strategies that provide data that can be analyzed and reviewed by their peers (other scientists). The U.S, Supreme Court ruled that natural science is testable against the natural world. Religion is not testable so cannot be taught in public schools because our constitution separates church from state.

"Can Creationism be taught in a Catholic School?" asked Noah.

"Yes, said Mrs. Marshall, "but I would be surprised since the Catholic Church accepts Darwinian evolution as being compatible with the Church.

Scientific Theory

Mrs. Marshall asked the class if they could explain a scientific theory reminding them that they just learned about the Theory of Evolution.

"A scientific theory must be supported by tons of evidence," said Caitlin.

"Right," said Mrs. Marshall. "In fact, any scientist worth his or her name in your school textbook accepts the idea of evolution."

"A scientific theory can be used to explain many observations," said Drew.

"Correct," said Mrs. Marshall.

"What about Scientific laws?" Asked Crystal.

"Glad you asked, Crystal." Mrs. Marshall continued, "In many ways, a scientific theory is more useful to scientists than a scientific law since it can explain a wide range of observations. Scientific Laws describe very specific events.

"Like the Law of Gravity," suggested Noah.

"Yes," said Mrs. Marshall.

"But Scientific Laws cannot change," said Sharde.

"Not exactly," said Mrs. Marshall. "Scientific laws can be extended or modified by a scientific theory. For example, Einstein's Theory of Gravity has been used to modify Newton's Law of Gravity because it better explains the motion of celestial objects. In fact, GPS depends on Einstein's theory not Newton for its accuracy. Now if your friend says, 'I have a theory as to why my teacher always wears sneakers everyday,' that is not a scientific theory. Some of the confusion about a theory comes from this type of use of the word. There is a big difference between a *theory* and a *scientific theory*.

Mrs. Marshall checked her watch and realized that it was time to leave even though there was more she wanted the children to see at the Science Center. "Time to go," she announced to her students. She and the students also thanked Mr. Hunter for all of his help.

"Too bad there is not time to see the remainder of the exhibit such as "Evolution is Good for Your Health," said Mr. Hunter.

"Perhaps, next time or we can do a virtual visit to the Science Center," replied Mrs. Marshall.

The students boarded the bus.

"What is the exhibit, 'Evolution is Good for Your Health' about?" inquired Drew.

Mrs. Marshall told her students she would talk about that exhibit in class the next day they met.

Evolution is Good for Our Health

Valves for the Heart

The next day, Drew reminded his teacher she was going to explain the exhibit on evolution and health.

"Does anyone here have a relative that received a new valve for his or her heart?" Asked Mrs. Marshall. Crystal raised her hand and said her grandfather had a new valve put in his heart. supports it has made possible many medical breakthroughs such as organ transplants and heart valves from pigs.

"My grandfather has a pig's valve in his heart! Said Crystal quite surprised.

Mrs. Marshall answered by saying, "Well you may not pick a pig as a close relative, but from an evolutionary viewpoint, they have a lot in common with us, saving many human's lives. Through surgery, pig heart valves have replaced the diseased or damage heart valves of millions of people. This is only possible because the structure of a pig valve is so similar to a human heart valve.

Diabetes

Next, Mrs. Marshall asked if her students knew anyone with diabetes.

A number of children raised their hands.

She explained to the students there are two main categories of diabetes, Type I, and Type II. Type II diabetes is related to obesity and more often occurs in adulthood. Type I diabetes occurs more frequently in children when their bodies can no longer make insulin.

Olive raised her hand and said her cousin has diabetes so it must be type I. Gavin raised his hand and said his grandmother was recently diagnosed with Type II diabetes

"What is insulin?" Asked Caitlin.

Mrs. Marshall answered the question by saying, "Insulin is a hormone needed to help your body move sugar into your cells so you can make the energy needed to live and do things from playing ball to reading a book. Because of evolution, the insulin produced by pigs, cows and sheep is very similar to human insulin and has been used to treat diabetes, saving many lives. However, since the insulin was not human insulin, there were some allergy problems. Scientists looked for a better treatment."

Lauren was raising her hand all excited. She said, "My cousin gets human insulin called Humulin. How is that possible? Do people donate insulin to her?

Mrs. Marshall replied. "Let me explain. Today, the human gene that regulates the production of insulin and other hormones such as the growth hormone can be placed into bacteria cells.

"Yuck, in bacteria cells," called out Sharde.

"I guess you are wondering how that is possible. As I have taught all of you, genes are composed of DNA. It does not matter whether you are tiny one-celled organisms, such as bacteria, a giant oak tree or a tiny mouse, the DNA in all living things is made of the same stuff. Because the makeup of the genes of all living things is essentially the same, bacteria have become factories for the production of important substances the human body cannot make on its own. You can say that scientists have "programmed" the bacteria to produce the needed substances. The human gene that carries the instructions on how to make human insulin is spliced into (placed into) the bacteria cells. Now the bacteria cells know how to make human insulin. And that they do it quite well allowing millions of people to lead normal, healthy lives!

"Wow, so my cousin gets her humulin from bacteria factories!" Called out Lauren. "That is soooo cool!"

Mrs. Marshall reminded Lauren to raise her hand if she has something to say though she appreciated Lauren's excitement over learning about humulin production.

Mrs. Marshall concluded by saying, "Heart valves and insulin are just two examples of how knowledge about evolution helps medical science.

"What happens if a doctor does not believe in evolution?" Asked Crystal.

"Funny, you should ask, Crystal, because I was just going to tell the class an interesting story about a doctor who did not believe in evolution. Mrs. Marshall told the class how in 1984, Dr. Leonard L. Bailey transplanted a baboon heart into the chest of Baby Fae, who was born with a severely defected heart.

"A baboon heart in a baby!" called out some of the students at the same time.

Mrs. Marshall showed the class a picture of baby Fae. She told them a graft between different animal species is called a xenograft. She wrote the name on the smart board next to the picture of baby Fae.

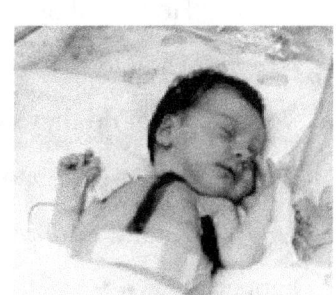
Baby Fae

"Did anyone ever have this type of graft before," asked Noah.

"No," replied Mrs. Marshall, "This was the first of its kind ever done.

"Sadly," she continued, "A few days after the graft, Baby Fae died because her body completely rejected the graft."

"Why did he use a baboon heart? You told us we are more closely related to chimpanzees," Asked Caitlin.

"Yes," replied Mrs. Marshall, "Doctors did question why he chose a baboon since according to evolution theory, it is not as closely related to humans as chimpanzees.

Mrs. Marshall then put up on the smart board an article from the *Times of London* that had published an interview between Dr. Bailey and an Australian radio crew.

She asked, Mike, another student in her class to read Dr. Bailey's reply to the question as to why he chose the heart of a baboon. Mike read Dr. Bailey's reply, "Er, I find that difficult to answer. You see, I don't believe in evolution."

Mrs. Marshall told the class that his response shocked the medical profession because he ignored such an important scientific theory.

"Do all doctors today believe in evolution?" asked Drew.

Most do, but I suspect there are some who choose to ignore all of the evidence, and as all of you well know, there is a huge amount of evidence supporting evolution," replied Mrs. Marshall.

CSI and Evolution: Gathering the Evidence

"How many of you watch some of the CSI TV shows?" Mrs. Marshall asked her class. Most of the students raised their hands.

"How do they solve crimes?"

"They look for evidence such as DNA, hair and fingerprints," replied Drew.

"What do they do next?"

"The investigators try to solve the crime by working with scientists," replied Noah.

"Right," said Mrs. Marshall. "They gather evidence and formulate hypotheses as to what happened."

Though Tanya was quiet during the Museum visit, she was very interested in this topic. She raised her hand and said, "Then, in many ways, the scientists who have gathered evidence to support the Theory of Evolution are sort of like crime scene investigators."

"Absolutely," replied Mrs. Marshall.

"So that giant crater found in the Gulf of Mexico provided evidence to support the hypothesis that an asteroid caused the extinction of dinosaurs," Tanya continued.

"That is correct, Tanya." Mrs. Marshall was pleased that her class was grasping the theory of evolution and decided to have a discussion with the class about other evidence supporting the Theory of Evolution.

Fossil Talk

Though they ran out of time at the Science Center, Mrs. Marshall arranged with the curator to do a virtual tour of the exhibits on evolution they did not have time to see. She projected on the smart board the exhibit called, 'Fossil Talk.'

Fossil Talk

Though they ran out of time at the Science Center, Mrs. Marshall arranged with the curator to do a virtual tour of the exhibits on evolution they did not have time to see. She projected on the smart board the exhibit called, 'Fossil Talk.'

"Who knows what a fossil is," she asked the class.

Jackson, a shy boy in the class, raised his hand slowly and said, "A fossil is anything that shows the remains of a living thing that no longer lives on this planet. I have some rocks with fossils of shellfish and trilobites at home."

"Perhaps you would like to share some of your fossils with the class," replied Mrs. Marshall.

Jackson nodded.

"How do Fossils 'talk' to scientists?' Asked Caitlin.

Noah raised his hand and said, "Well just finding a fossil of an extinct animal tells you that it once lived on Earth. That is sort of like talking to scientists."

Mrs. Marshall nodded while saying, "Obviously, 'Fossil Talk' is a metaphor for what information scientists are able to extract from fossils."

T-Rex

Evolution of the Web

http://www.enchantedlearning.com/subjects/dinosaurs/

Next, she brought up on the smart board the dinosaur exhibit showing dinosaur skeletons and eggs. She asked Sharde to read what was written on the Board.

"Almost complete dinosaur skeletons have been put together from these bones. The neat thing about these skeletons, once they are assembled, is that they 'talk' to scientists. They can tell them how they lived and what they may have eaten."

Mrs. Marshall then put on the White Board an exhibit about human evolution. She told the class there is fossil evidence showing other types of humans lived on the planet. A picture of a Neanderthal skull appeared on the Board. She told the class that though Neanderthal no longer lives on Earth, it is evidence that other types of hominids (humans or relatives of humans) lived on Earth such as Homo erectus and Homo habilis." She told the class the Science Center will soon have an exhibit on human evolution.

Mrs. Marshall, displayed on the white board an exhibit called "Missing Links"

"Caitlin raised her hand, "Isn't a missing link an animal that connects an extinct one to a modern day one?'

"Yes, replied Mrs. Marshall. Missing links are transitional species and provide more evidence for the Theory of Evolution because they show how one organism evolved into another organism."

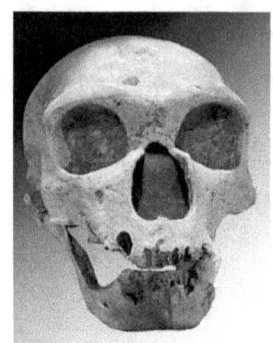

Neanderthal skull

"Isn't there a bird-like dinosaur that support the idea dinosaurs evolved into birds as Mr. Hunter told us yesterday," said Noah.

Mrs. Marshall then showed a picture of how scientists think the Archaeopteryx looked and said they it is a link between dinosaurs and birds because it had characteristics of both. It had feathers like a bird but teeth and a bony tail like dinosaurs.

She then showed the students a video from the Science Center about another missing link called Tiktaalik, a fossil discovered in 2004 that links land to water animals.

Archaeopteryx

Evolution on the Web
Check out the discovery of the Tiktaalik in this video:
http://www.pbs.org/wgbh/evolution/library/03/4/l_034_49.html

Tiktaalik

Mrs. Marshall continued the virtual tour through the Science Center that provided additional evidence supporting the Theory of Evolution.

"I hope all of you go back to the Science Center and visit the exhibits we missed. If there is money in the school budget, perhaps we can go back as a class. However, our discussion of evolution would not be complete without providing you with some of the other importance evidence supporting this theory."

Next, Mrs. Marshall placed on the white board the exhibit titled, "Embryology: From Egg to Birth or Hatching."

Life is A'Changin', **Tales of Science** by Joan Wagner

Embryology: From Egg to Birth or Hatching

Drew raised his hand and said, "Doesn't embryology study how living things develop?"

"Yes," replied Mrs. Marshall, "It studies the development of living things from the fertilization of an egg to its birth or hatching."

Caitlin raised her hand and said "My mom showed me a picture of me when I was only a few weeks old and then when I was 3 months old. I really changed."

"Undoubtedly, Caitlin looked quite different when she was a one-month old embryo inside of your mom compared to 3 months," replied Mrs. Marshall.

Mrs. Marshall soon learned that many students in her class have seen ultrasound pictures of themselves so she asked them, "Now what do you have in common with a fish, pig and chicken? Not sure? Then she showed them this video from NOVA: https://www.pbslearningmedia.org/resource/tdc02.sci.life.cyc.embryo/common-past-different-paths/

Mrs. Marshall watched the expression on her students' faces as they watched the short video being projected from the Science Center.

And then said, "Pretty amazing!"

She asked Drew to read what was being projected on the white board.

"This resemblance is not surprising to evolutionary scientists. After all, why should nature get rid of a good body plan? Instead, it has been preserved over and over again in living things. As you can see, all of these animals, as embryos look remarkably alike. This is not a coincidence. It is part of Darwinian *Natural Selection*. The body plan had survival value. Of course, after birth or hatching, these animals lose a lot of their resemblance, but the basic body plan is still there. They all still have a head with two eyes, body trunk and a tail region. In humans your tail has become your coccyx bone (what you sit on).

"What about homologous structures?" called out Noah when he noticed that listed on the white board.

"Homologous structures provide more evidence for evolution," said Mrs. Marshall.

Homologous structures: It's all in the bones

Mrs. Marshall asked Noah to read what the Museum exhibit says about homologous structures.

"Now, what do you have in common with a cow, horse, whale and bird? She then showed them this video on homologous structures: https://www.ck12.org/flx/render/embeddedobject/157380.

"As you must see from this video, the skeleton of your arm is amazingly similar to those of these animals. Scientists call these structures *homologous* because their anatomy provides evidence for a common ancestor. If they do not share common ancestry, then why are these structures so similar? Homologous structures provide powerful evidence for evolution because they show how living things are related to one another."

"What about these analogous structures," asked Leticia, another student in the class. Mrs. Marshall had projected the analogous structures from the Science Center exhibit. Since Leticia asked the question, she called on her to read from the exhibit.

Analogous Structures

"What does a butterfly, bird and bat have in common?"

"Okay, you guessed it! They all can fly. Obviously, the development of flight had survival value to both animals. However, if you look at the anatomy of a butterfly wing and compare it to a bird, the two are quite different. They developed independently of one another, not sharing a common ancestor for the structure. Since the structures have the same function (both used for flight), the structures are said to be analogous. Bats are flying mammals yet their ancestors did not fly. They are an example of a mammal whose forelegs became modified into wings. This was an adaptation that had survival value for them. They could better escape enemies, hide out in caves and hunt flying animals. Their wings are also analogous to a butterfly because they have the same function. Like the evidence in a crime scene, analogous structures tell scientists that living things have changed to adapt to their environment."

Mrs. Marshall then showed this video from the Science Center about analogous structures.\: https://www.ck12.org/flx/render/embeddedobject/157381

"That video is cool," said Leticia.

Any questions? Asked Mrs. Marshall.

Jackson raised his arm and then said, "So can we say bird wings are homologous to our arms but analogous to the wings of butterflies?"

"Right," replied Mrs. Marshall, smiling because her class was really grasping the topic of evolution and the evidence supporting the theory.

"I had my appendix removed. My mother said I did not need the appendix because it is a vestigial structure. Does that have anything to do with evolution?" Asked Leticia.

"I am glad you brought that up because that is another example of evolution." Said Mrs. Marshall. We have parts we no longer use and have been reduced in size."

"Like our coccyx bone, which is a left-over tail," called out Noah.

Mrs. Marshall then showed the video on vestigial structures to her class: https://www.ck12.org/flx/render/embeddedobject/157379.

This diagram summarizes homologous and analogous structures: Reading vertically are analogous structures, a cat's leg compared to a praying mantis leg and a whale flipper compared to the insect called a water boatman. Reading horizontally, a cat's leg is compared to a whale flipper and a praying mantis leg is compared to the leg of a water boatman.

	Analogous Leg	Analogous Flipper
Homologous: Mammals	Cat Leg	Whale Flipper
Homologous: Insects	Preying Mantis Leg	Water Boatman Flipper Leg

The last evidence to support evolution being projected from the Science Center was DNA.

It's All in the DNA

Noah raised his hand and said, "We already learned the DNA that makes up our genes is the same in all living things so DNA is DNA whether you are a mouse or an elephant."

Lauren then said, "My cousin gets Humulin made in bacteria because the human insulin gene can be placed or spliced into bacteria DNA resulting the bacteria becoming a human insulin "factory."

Mrs. Marshall clarified a bit by saying "Yes, the basic structure of human DNA and all other forms of life is identical. DNA is the substance that makes up your genes. Genes provide you with all of your traits. Even, the lowly fruit fly shares some of the genes found in humans!

DNA can also talk to scientists," said Mrs. Marshall. You certainly have seen it on the CSI shows. A scientist uses DNA to identify a person or any living thing or to show similiarities among living things."

Mrs. Marshall asked Caitlin to read what the exhibit concludes about DNA.

"DNA provides scientists with some of the strongest evidence for evolution. When DNA changes (mutates), new traits may appear. If these traits have survival value to an organism they are kept and passed down to their offspring. You can say that DNA is the "engine" of evolution. When the DNA in the coding part of a gene mutates, life changes. If the change improves survival, the organism evolves.

Mrs. Marshall then projected the last statement in the exhibit on evolution.

Evolution is the scientific theory that living things change, and those changes with survival value help living things to adapt to their environment. The English naturalist, Charles Darwin proposed this theory back in the 1800's. Scientists are still trying to understand more about what causes evolution, but they all agree that living things do indeed change. Understanding evolution will lead to cures and treatments for diseases. It is one of the reasons we need to do something about climate change because there is substantial evidence the climate is changing faster than living things can adapt, threatening their very existence. It is the reason we all need to become good stewards of the environment (protect the environment).

Mrs. Marshall then said to her class, "I want to leave you with this thought as we journeyed through the science of evolution. I hope understanding evolution will give you a better appreciation of all life, for we are all connected to one another. In humans, hair, skin and eye color are superficial variations among people. All humans are much more alike than different.

Discussion Questions

1. The dinosaurs went extinct about 65 million years ago. What is the theory that explain the extinction and what is the evidence for the theory?
2. Today. Many scientists believe birds evolved from dinosaurs. What is the evidence that that led scientists to this conclusion?
3. How does the fossil record tell when organisms lived?
4. Who was Charles Darwin and what was his contribution to our understanding of life on our planet?
5. How did Darwin's finches adapt to the environment?
6. What did Darwin mean by "Survival of the Fittest?"
7. How do antibiotic resistant bacteria support Darwin's theory of evolution?
8. Why did Darwin's books cause controversy?
9. What is Creationism and Intelligent Design?
10. What is a scientific theory?
11. How has our knowledge of evolution been helpful with organ transplants? What were some of the issues regarding the heart transplant for Baby Fae.
12. What are some examples of how the fossil record supports evolution?

Life is A'Changin Science Terms

1. **Adaptation**: Ability of organisms to change for survival
2. **Analogous structures**: Structures on organisms that have the same function but are structurally different such as insect wings and bird wings.
3. **Asteroid**: A celestial rock orbiting mostly between Mars and Jupiter.
4. **Charles Darwin**: Developed theory of evolution often called "Survival of the Fittest
5. **Darwin's finches**: Darwin used the variability of finch beaks to develop his theory of evolution
6. **Diabetes**: A disease in which too much sugar accumulates in the blood.
7. **Dinosaur**: A very successful group of organisms that went extinct about 65 million years ago.
8. **DNA**: The chemical substance that carries traits. Most genes are made of DNA.
9. **Embryology**: The study of the development of organisms that has been also used to support evolution because it shows similarities in development for many organisms
10. **Evolution**: The theory that life on Earth is changing
11. **Extinct**: When an organism disappears from Earth
12. **Fossil**: Evidence of past life from bones, rock impressions or stone casts of living things.
13. **Homologous structures**: Structures of living things that developed the same way such as the wing of a bird and the fore leg of a dog.
14. **Humulin**: Genetically engineered insulin
15. **Impact ejecta**: Rock emitted after being struct by an asteroid
16. **Iridium**: A mineral found in abundance on asteroids and comets but rare on Earth and has been used to support the extinction of dinosaurs by an asteroid.
17. **K-T Boundary**: The geologic boundary between the Cretaceous and Tertiary geologic periods.
18. **Radioactive dating**: A method used to date fossils using the decay rate of radioactive minerals associated with fossils.
19. **Relative dating**: A method that uses the layer a fossil is in to determine its relative age to other fossils found in the same rock.

20. **Resistant bacteria**: Bacteria that evolved to be resistant to antibiotics.
21. **Shocked quartz**: Quartz melted and then solidified after being placed under great pressure, such as being struck by an asteroid.
22. **Survival of the Fittest**: Darwin's theory of evolution that living things best adapted to the environment survive.
23. **Variations:** All living things display variations. And those with variations beneficial to survival pass on those traits to their offspring.
24. **Vestigial organ:** Structure left reduced in size and not longer has a function for a living organism such as the appendix.
25. **Volcano**: A mountain that can release lava as the tectonic plates move.
26. **Xenograft:** A transplant between two different organisms such as the baboon heart given to Baby Fae.

NGSS Standards

MS-LS4 Biological Evolution: Unity and Diversity
Disciplinary Core Ideas
Evidence of Common Ancestry and Diversity
1. The collection of fossils and their placement in chronological order (e.g., through the location of the sedimentary layers in which they are found or through radioactive dating) is known as the fossil record. It documents the existence, diversity, extinction, and change of many life forms throughout the history of life on Earth.
2. Anatomical similarities and differences between various organisms living today and between them and organisms in the fossil record, enable the reconstruction of evolutionary history and the inference of lines of evolutionary descent.
3. Comparison of the embryological development of different species also reveals similarities that show relationships not evident in the fully-formed anatomy.

Natural Selection
1. Natural selection leads to the predominance of certain traits in a population, and the suppression of others.
2. In artificial selection, humans have the capacity to influence certain characteristics of organisms by selective breeding. One can choose desired parental traits determined by genes, which are then passed on to offspring.

Adaptation
1. Adaptation by natural selection acting over generations is one important process by which species change over time in response to changes in environmental conditions. Traits that support successful survival and reproduction in the new environment become more common; those that do not, become less common. Thus, the distribution of traits in a population changes.

Science and Engineering Practices
- Analyzing and interpreting data
- Obtaining, evaluating and communicating ideas

Crosscutting
- Patterns
- Cause and effect
- Interdependence of science, engineering and technology

Project Pond/Wetland

It was the first day of school and Mr. Hull was very excited about utilizing the pond, wetland, butterfly field students, faculty and the community built for the middle school. He and other science teachers in the building had collaborated on the development of an extensive curriculum for the middle school during the summer. The students at Brownville Middle School were as excited as the teachers.

Savannah, Madison, Jayden and Owen were four of the students that had helped to organize student labor for the construction of the pond the previous year. They were 8th graders now, so this would be their first and last year to interact directly with the Pond Project/Wetland in school. They were all in Mr. Hull's science class.

"Welcome class. I hope all of you had a wonderful and productive summer vacation. I gather you have noticed the pond outside of the window. I know many of you helped with this project last year and even came in during the summer to help," said Mr. Hull, welcoming his students.

"Can we go out to the courtyard and visit the pond?" asked Savannah.

"Absolutely," replied Mr. Hull.

Once he had taken attendance, assigned seats and distributed books, he took the class out to the courtyard.

The students circled around the pond and wetland first, looking around and pointing to things that interested them.

"The pond/wetland is an example of an ecosystem," said Mr. Hull. "Who knows what an ecosystem is?

"It is how living things interact with one another and their non-living environment," said Madison "Look, a painted turtle!"

Jayden's eyes were glued to another section of the pond. "I think we are going to have some fish babies," he observed. The students turned and saw two fish mating on the sandy bottom of the pond. "Those are sunfish," Jayden added. "Soon there will be many more."

"Some more," corrected Savannah. "Some, will be eaten by the larger fish. And that is a good thing or else the pond will not have enough room or food for the other forms of life we placed in the pond.

"There goes a newt," said Owen, pointing to one on a rock.

"What happened to all of the frogs?" Asked Savannah to the teacher.

"I am not sure," said Mr. Hull.

"But I put five of them in the water during the summer," replied Savannah.

"Well we now have a mystery to solve," said Mr. Hull.

Madison admired the water lilies that were placed in pots in the bottom of the pond. One of the rules set for the pond is all life placed in it must be indigenous to the School's region.

"Remember when Mrs. Andrews brought in a beautiful purple flower to place in the bottom and you told her it cannot be placed there because it is an invasive species," Madison said to Mr. Hull.

"Do you want to explain to the class what is meant by an invasive species? Mr. Hull asked Madison.

"It was really funny," said Madison. "We were all stocking the pond with animal life and in walks Mrs. Andrews with the purple flowering plant. I found this pretty wetland purple flowered plant and I thought it would work well in the pond," she said. "I loved the expression on your face Mr. Hull."

"Yes, I did have to tell Mrs. Andrews that the plant was called purple loosestrife and it was an invasive plant. Can you explain what I meant?"

Madison said, "An invasive plant is not from this area and when it is introduced it interferes with the survival of other organisms because there are no other organisms to keep it in check. Purple loosestrife can reproduce so rapidly that other wetland plants such as cattails become threatened."

"And why else do you think that is a problem?" Mr. Hull asked of the class.

Wikipedia.com

Owen raised his hand. "Because different animals eat different plants. The food of some animals may not be available and then those animals may die out and animals that eat those animals will die and so on."

Mr. Hull thought this was a good opportunity for a fun assignment. "I would like you to work in a team of 2-3 and develop a model for the food relationships in the pond. I will give you time in class to work on this. I would then like you to produce a second model that shows the flow of energy in this pond environment.

"I see something swimming but I can't figure out what it is," said Sara, another student in the class.

Mr. Hull distributed some small collection vials to the students and asked them to take some samples of the pond water so they could view the samples through a microscope the next day. "Tomorrow, Sara, you and the rest of the class will learn about organisms you can barely see and some that you cannot see without a microscope."

"Jayden, can you explain to the class what a wetland is and how it is beneficial to an environment?" asked Mr. Hull. He knew Jayden was educated on this from working on the pond over summer.

Jayden thought for a moment and then remembered the discussion when they were putting in the wetland. "A wetland is an area that stays wet a certain amount of time each year. The area off the pond is a wetland. It collects runoff from the pond. The cattail we planted in it is a common wetland plant in our region. Wetlands provide homes to many living things and also can control flooding," he explained, shooing away a dragonfly.

"Can't we consider the pond a wetland too?" asked Madison.

"Yes," replied Mr. Hull, "but the area we have designated as a wetland also controls flooding from the pond when we have a lot of rainfall."

"Let's take a walk over to the butterfly field," said Mr. Hull. The students followed him over to another section of the courtyard.

"Look, a chrysalis, "called out, Owen.

The students stopped to observe. Suddenly the chrysalis started to vibrate. The students' eyes were all glued to the chrysalis as it hung off the school building. In a few minutes a monarch butterfly emerged, spreading out its wings. Mr. Hull walked over to it and allowed the butterfly to crawl onto his fingers.

"Mrs. Nielson's 7th grade class is going to tag the monarch butterflies to monitor how far they migrate. It is part of a national program coordinated on a college campus to learn more about their migration patterns. Each tag has a number and when the butterfly is captured the number will be placed on the Internet with the location in which it is found."

"I saw the butterfly tent Mrs. Nielsen set up in her classroom to house and tag the butterflies," added Madison.

"How do they tag a butterfly?" asked John, another student in the class.

"The tags are made to stick to the wing of the butterfly without harming it. I can show you when we go inside," offered Mr. Hull.

That is so cool that we can actually release a butterfly and track where they go," Savannah said, smiling to her friend.

The children walked over to where the milkweed was planted. A number of monarch caterpillars were found on the underside of the

MONARCH WATCH TAG

leaves. However, what impressed all of them was the number of chrysalises they found throughout the courtyard.

"It just shows that when plenty of food is available for a species, they can really flourish," Madison said to Savannah. "I hope there is lots of milkweed for the butterflies when they migrate."

"I heard that many people think milkweed is an undesirable weed and so they get rid of it," said Owen.

"That is not good news for the Monarch," Savannah added sadly.

Mr. Hull heard his students talking and asked them, "How can we help the Monarch?"

"We need people to become aware of the fact that milkweed is very important to the survival of the Monarch butterfly," suggested Madison.

"We can plant milkweed in our backyards and get permission to plant some in the parks," Owen suggested.

"Don't forget," said Jayden, "we also have to plant an assortment of flowers like we have in our courtyard to provide nectar to the adult butterflies. The milkweed is food for the caterpillars or larva but not for the adults."

"We do have a pretty flower garden here," noted Owen, smiling.

Mr. Hull told the students the project has been so successful that many teachers, administrators and the custodial staff were looking to put in ponds and have a butterfly field.

Mr. Hull checked his watch and told the students to head back to the classroom since the period was going to end shortly.

The students who collected samples of water brought them into the classroom.

"Better open up the vial to allow oxygen to get in," suggested Madison to her friend Sara, who was carrying one of the vials.

Just as she finished giving her suggestion, Mr. Hull told all students carrying vials to loosen the lids to allow oxygen in, but keep them covered so they do not get contaminated with particles in the air of the room.

Mr. Hull reminded his students their assignment was to prepare a model for the food interactions going on in the courtyard and a model that demonstrated the flow of energy. The project was due the following Monday. The students were allowed to work in teams of up to 3.

Madison raised her hand and said, "Mr. Hull, I understand how to tackle a food relationship model, but I am not sure what you mean by 'flow of energy'."

From the look on everyone's face in the classroom, Mr. Hull realized this topic needed more explaining. There were still a few minutes left and he explained, "All living things need energy. Some living things are able to make their own food, such as plants while other living things get their food from another organism. That organism can be dead or alive."

Madison raised her hand. "I know plants get their energy directly from the Sun."

"I get my energy from the food I eat such as chicken or carrots," said Jayden.

"Do any organisms get their energy from dead things?" asked Mr. Hull.

Gibrail raised his hand and said, "When plants or animals die, they decay because organisms like bacteria are eating them."

"Excellent explanations," said Mr. Hull. "Energy that comes from the Sun in a pond or wetland ecosystem gets transferred from one organism to another. In addition to food relationships, this is also what I want you to model. Are there any other questions? If you cannot think of one now, when you are working on this project in class, I will be available to answer questions you may still have on how to model energy transfers."

The bell rang and the students proceeded to their next period class.

"We cannot prepare a model of the food interactions unless we know all the main organisms that are out in the courtyard," Savannah pointed out to her partner Madison, while walking to their next class. Their friends Jayden and Owen also formed a team. Most of the students in their class formed teams though some chose to work independently.

"Let's stay after school and survey the pond," suggested Savannah.

"OK. See you in Mr. Hull's room after school.

"Count us in," yelled Jayden and Owen at the same time.

As planned, they all met in their science classroom during the school's after school activity period and Mr. Hull let them into the courtyard. Before going out there, the two groups took a few copies of a pond life guide to help them identify some of the organisms.

"First we need to survey the life that is out here and then we need to decide what they eat," said Madison.

"There is a painted turtle," Jayden told them a few minutes later after he had spotted one.

"No, I think it may be the slider turtle though I know the pond has both. They look very similar. But because they are so similar and eat the same food, it will not affect how we design our model of the food relationships," said Savannah.

Since they did not have a microscope with them, they could only survey what they could see. These are organisms they found in and around the pond/wetland.

1. Sunfish
2. Minnows
3. Newts
4. Salamander
5. Algae
6. Waterlilies
7. Dragonflies
8. Mosquitoes
9. Mayflies
10. Crane flies
11. Water striders

12. Tadpoles
13. Painted turtle
14. Slider Turtle
15. Cattails

"I think some of those small things we see swimming may be the larva of some of the insects. We should be able to see them better tomorrow when we look at some of the pond water under the microscope," observed Jayden.

"I sure would like to know what happened to the frogs," Madison said sadly. "I will bring some more to put in the pond and then we can add them to our model."

They all said their goodbyes and looked forward to working on their models.

The next day, as promised, Madison brought in 4 frogs she captured in a creek that ran next to the school. "Here, Mr. Hull, I have some frogs to add to the pond."

"Thanks, Madison. Here is the key to the courtyard. You can add them to the pond now," said Mr. Hull.

Madison carried the frogs out to the pond, placed them in the water and watched as they swam away.

"They should have a happy home here," she thought.

During science class that day, the students looked at the pond water under microscopes. They found flatworms, Paramecium, Amoeba, Euglena and the larva of a number of insects. Mr. Hull had a number of guidebooks to help them identify the organisms. The class time went by quickly as they all surveyed the pond for microscopic life. Some teachers passing by the lab walked in to look through the microscopes.

"Now our models for food interaction must include these small forms of life too," noted Jayden.

Mr. Hull told the students that it was their call as to what to include. "Remember, a model explains what is happening. You decide how detailed of a model you want to construct. All teams will present their model to the class and explain what is happening next week."

All of the models were due on Monday of the following week. The students had signed up for when they would present their model to the class.

The next day, they all worked on their models during class. As they were working, they could hear an echo of frogs croaking from the filter in the pond.

Savannah raised her hand and said, "It sounds like the frogs are inside the filter."

Mr. Hull sent her out to the pond to check. Sure enough, the frogs were all in the filter. Savannah took them out and placed them back into the pond before she went back to class.

About 10 minutes later, they all heard the same sound again.

At that moment, Mrs. Nielsen stopped by to say they had already tagged 10 monarchs.

Mr. Hull asked her, "Can you figure out why the frogs keep going into the filter?"

"Easy," she replied. "They are trying to get away from the turtles. Turtles do like to eat frogs when given the opportunity and this enclosed space provides a great opportunity for them."

Owen raised his hand and said, "I think we now have solved the case of the missing frogs. They were dinner for the turtles."

The rest of the class took quick notes to add to their models of food relationships in the pond.

Over the weekend, they decided to meet at Madison's place to put the finishing touches on their models. Madison and Savannah did their models on poster paper. Jayden and Owen decided to do three-dimensional models using string and paper.

Jayden said, "Let's cut out pictures of all of the organisms we surveyed in the pond and paste them to index cards. Next we can punch holes in the card and use string to attach one organism card to an organism card that provides it with food."

"Great idea," said Owen.

"We better research what eats what before we attach everything together," noted Jayden.

Soon Jayden and Owen were busy assembling their food interaction model. They were careful not to tangle up the yarn they used to attach the cards. Owen's mom used to crochet so she was happy to donate yarn to be used as the string in the project.

Meanwhile, Madison and Savannah were busy assembling their version of food relationships in a pond. They too did some research to determine who eats what.

"Do you think we have too many arrows on our poster?" asked Madison.

"Well, we should show all of the relationships," replied Savannah.

"But a model doesn't have to show everything, it just has to provide a general idea as to what is happening," stated Madison

"I disagree," said Savannah, "I think we should provide as much information as we can in the space we have."

"But it will look messy," said Madison.

"This is not an art project," said Savannah. "We are trying to model all of the interactions in a pond ecosystem. The more interactions we provide, the more useful it will be."

"I guess you are right," Madison sighed. "But we should do our best to make it as neat as possible so that the model can be understood, otherwise, it will be of no use to anyone."

After attaching pictures of the forms of life onto the poster and drawing arrows, they sat back to admire their model and decide whether it was complete.

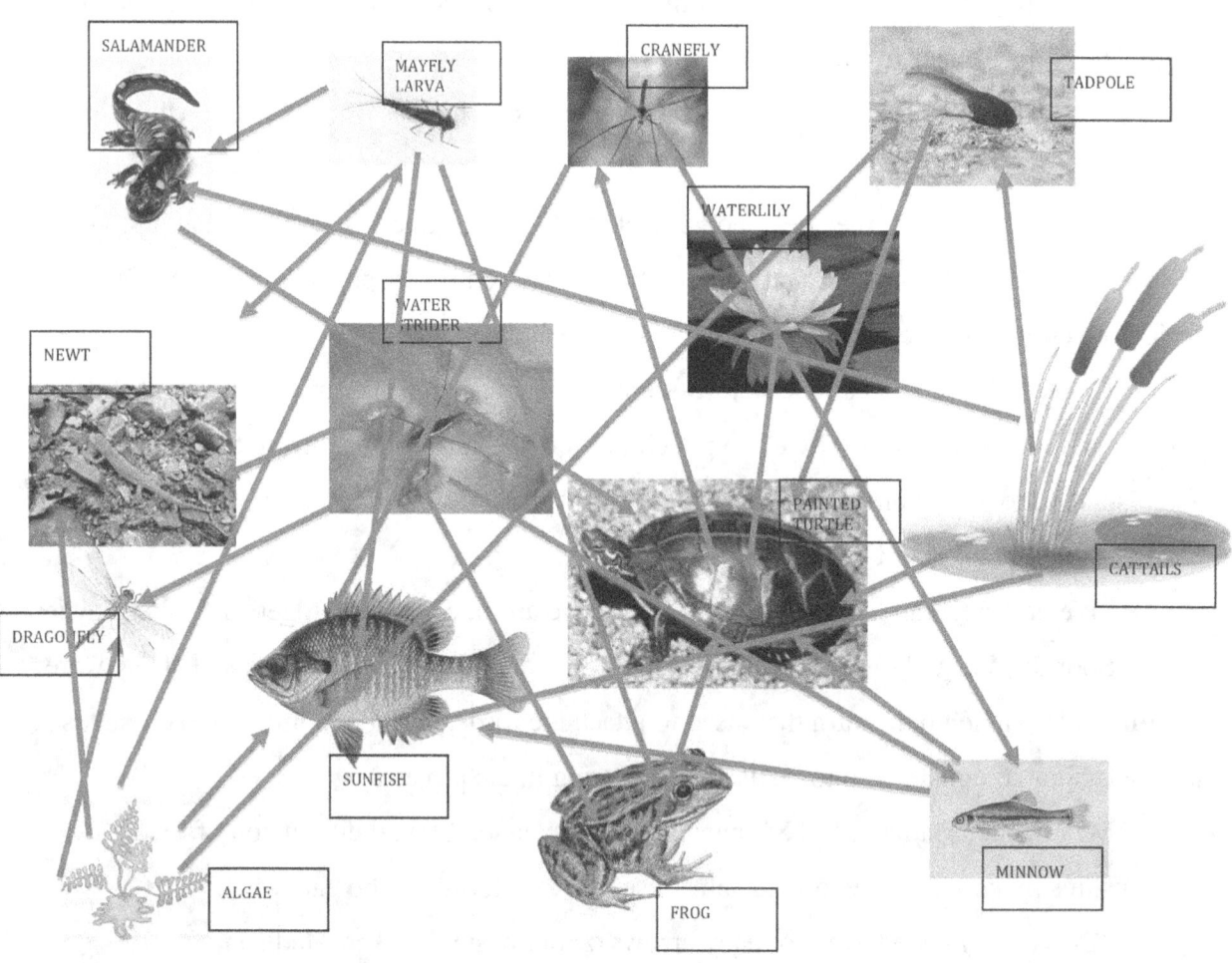

"Wow, it looks like the turtle is the main consumer in this model. Now I know why they call food relationships in an ecosystem, food webs. This sure looks like a giant web," stated Madison after examining their project.

"I think this is a fairly good representation of what is happening in a pond. Of course, we did not include all of the microscopic life, but I think this is a fairly good representation of what is happening, "said Savannah.

"Something is missing," stated Madison.

When they walked over to Jayden and Owen's model, they noticed what they forgot to include.

Savannah found a picture of decay bacteria and included them in the model to show that when organisms die, their nutrients are returned to the soil and then used by plants to make food.

"We should include a picture of the sun to show that plants get energy from the sun to make food," Madison noted.

Madison and Savannah got to work to make these additions but they used the sun picture to illustrate energy transfers in their food pyramid picture.

They looked at their model again and decided it told the story about food relationships in an environment.

"Well the turtle still stays the main consumer in this model," said Savannah.

"Do you think the dead fish is enough to show how organisms get recycled in nature?" asked Madison.

"Yes, "replied Savannah. "It is enough to illustrate the point."

Savannah and Madison admired their work and then set out to show how energy moves within a pond environment.

"I think the best way to represent this is with a pyramid. I have seen this done before," stated Savannah.

"Since the base of a pyramid occupies the greatest area, the organisms that provide the most energy should be represented here," stated Madison.

"Obviously, that would be the plants or any organism that can make their own food," Savannah said.

The girls decided to place the algae, cattails and water lilies at the bottom of the pyramid.

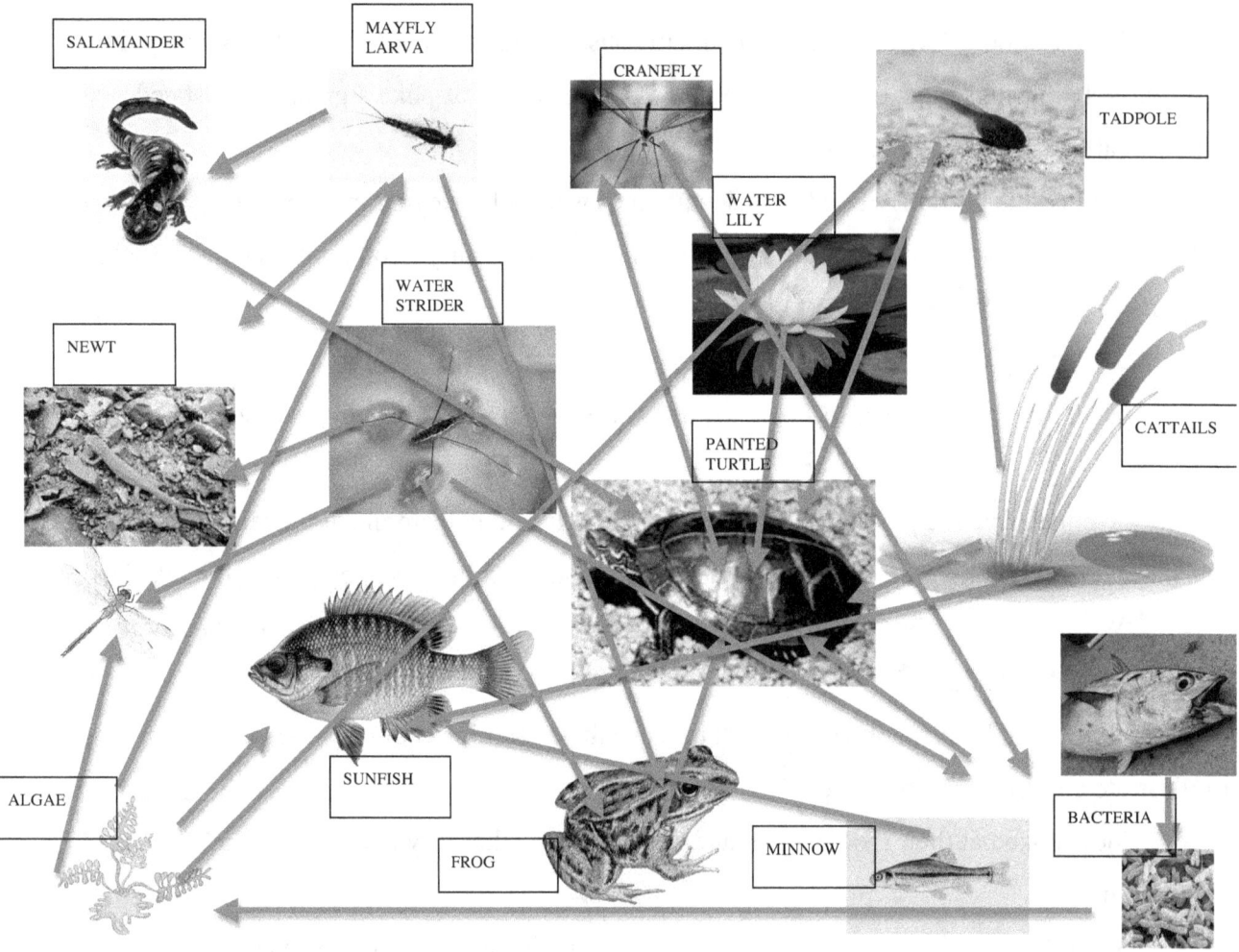

"As we go up the pyramid, there is less and less energy that can be transferred, because the population of those organisms are less," said Madison.

Owen and Jayden overheard the girls talking. Jayden said, "Of course there has to be more of the organisms that provide the food for another organism or soon there would not be enough food for anyone in an ecosystem."

"Yeah," agreed Owen. "If all of the insects in the pond suddenly died, the organisms that use them for food would go hungry and die or might begin eating food that's normally used by another organism and there would be a big mess!"

"Yes, this is clearly what is meant by maintaining a balance of nature. There needs to be enough food for everyone or the balance gets disturbed and some organisms may even go extinct," concluded Madison.

"That is a depressing thought," Jayden said.

Project Pond/Wetland, **Tales of Science** by Joan S. Wagner

The two teams went to work on creating their models of energy flow.

Jayden and Owen stuck with their 3D plan. They cut out bigger and bigger triangles and linked them together with string. Savannah and Madison drew their model on poster paper.

"The algae, water lilies and cattails should be at the base of the triangle because they make their own food while the turtle should be on the top since it seems to eat a lot of all the other organisms," Savannah observed.

"I agree," replied Madison. We just need to decide where to place all the other pond organisms."

"I think the criteria we should use are their size and populations. The more they are represented in the pond ecosystem, the lower on the pyramid they go because they are providing more energy than the higher tiers on the triangle."

Madison and Savannah first drew a rough draft of their model showing energy exchanges and then they created the final poster to submit to Mr. Hull's class.

Savannah and Madison sat back and admired their poster before showing it to Jayden and Owen.

Owen commented, "Looks good, but shouldn't you place the tadpoles with the frogs and salamanders?"

"Nope! Frogs always lay a lot more eggs that hatch into tadpoles, but most of the tadpoles are eaten by fish and even frogs so the population remains stable," replied Savannah.

"Hmm, I get your point," Owen replied. "I think we have some corrections to make on our model, Jayden."

The two boys went back to their model and with a little cutting of yarn, they made a better representation of energy changes in the pond ecosystem.

On Friday, all of Mr. Hull's students presented their models of food relationships and energy exchanges. Some students prepared relatively simple models showing only one source of food per organism. These models were called food chain models.

Sara and Alba's model used food chains.

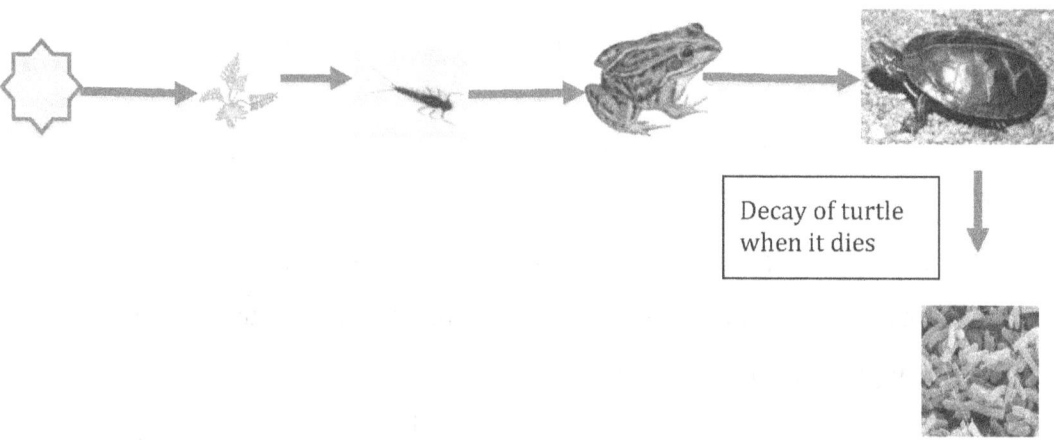

Madison and Savannah were very excited about presenting their models to the classroom. It was decided that Savannah would present the food web and Madison would present the energy changes. Each would hold up the poster being discussed for the other person. One member of the team held up the poster while the other person discussed it.

Savannah explained to the class: "As you can see, our model (using the picture that showed decay bacteria) shows that the food relationships in a pond are fairly complex. The important thing is to have plant and plant-like organisms that get their energy from the Sun in an ecosystem. The water lilies, cattails and algae provide the primary food for all life in the pond/wetland. They are also called producers because they make their own food. As you can see in our model, many of the organisms eat the producers. It seems the rule in most ecosystems is the big eat the small. Many of the pond organisms eat insects such as the fish, salamanders and frogs. You will notice that the turtle seems to eat almost everything it can. Normally, it is hard for it to eat frogs, but the mystery of the disappearing frogs was solved when we learned that turtles will eat frogs when given the opportunity. The poor frogs we put in the pond would try to hide from the turtles in the pond filter. I guess we were not doing the frogs a favor when we put them back in the pond. Obviously, the turtles found them yummy. I learned that the population of each organism remains stable as long as they have enough food and shelter in which to live."

Mr. Hull complimented Savannah on her presentation and asked the class if there were any questions.

Mike raised his hand. "How come you did not include the milfoil weed that has contaminated the pond?"

Savannah replied, "We could have included it, but Madison and I are working hard to get rid of the weed since it will upset the balance of nature in the pond. You must have noticed we are outside almost every day removing the weed with a net. If it got out of control, many of the pond life would be threatened because of lack of oxygen and the destruction of food sources."

Mike asked, "How can there be a lack of oxygen, when plants release oxygen?"

Madison decided to take that question. She remembered the discussion she had with her teacher about the need to rid of the milfoil weed earlier. "True, milfoil releases oxygen during photosynthesis, as do all plants, but when it dies, bacteria use oxygen to decay it. Because of this, more oxygen will be removed than is put in by the plant. The living things in our food web need oxygen or they will die."

"I see," said Mike a bit impressed by Madison's knowledge.

When it was Madison's turn to explain the energy relationships, she pointed to the poster and said, "We decided to represent the energy relationships with a pyramid shape. Organisms on the bottom of the pyramid provide most of the energy. As you can see, these are the producers.

Next are the primary consumers, the ones that eat the plants. These consist mostly of insects though some of the other animals eat both plants and animals. The animal on the top of the pyramid has the least amount of energy transferred since it eats most of the organisms. We had trouble deciding whether to place the fish, frogs, salamanders and newts on the same tier of the pyramid. We decided there are probably more fish than the amphibians so we placed them higher up on the pyramid. The Sun is the source of all energy for a pond system either directly or indirectly.

For example, the energy the turtle gets from eating a tadpole is traced to the energy the tadpole obtains from eating an insect, which is traced to the energy an insect gets from eating a plant, which is from the Sun.

Owen and Jaden gave a similar presentation holding up their mobiles of food and energy relationships.

At the end of the class, Mr. Hull complimented the students on their presentations and hoped they now appreciated the usefulness of modeling observations in the environment. He then prepared them for the next pond/wetland project.

He asked the students to bring in samples of water from local ponds, lakes, creeks, and other bodies of water since they would be testing the quality of the water and comparing it to the quality of the pond water. He told them the science of water is called limnology.

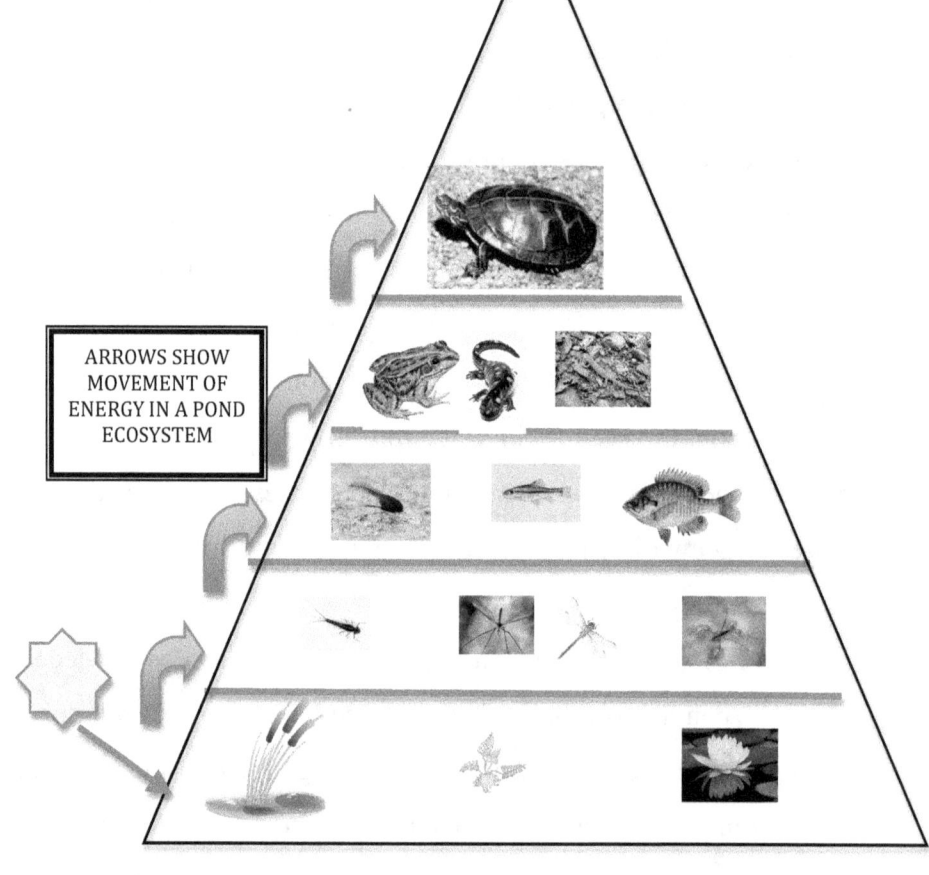

ARROWS SHOW MOVEMENT OF ENERGY IN A POND ECOSYSTEM

He then wrote on the whiteboard, the following tests they would be performing.

- Dissolved oxygen,
- Carbon dioxide,
- Ammonia-Nitrogen,
- Nitrogen-Nitrate
- Chloride
- pH
- Iron
- Phosphorus

Owen raised his hand after the list of tests was written on the board. "Remember when the pH of the pond was up to 12 during the summer?" asked Owen.

"Yes," replied Mr. Hull, "and you were very helpful to test the water each day you came to work on the pond during the summer.

"Do you remember how we solved the problem?" asked his teacher.

"Yes," said Owen. "We needed to lower the pH. We did this by running the filter longer each day. The filter was on a timer so it did not run all of the time. When we increased the time it ran, more carbon dioxide was able to dissolve in the water. When it dissolves it forms a weak acid, called carbonic acid. This helped to lower the pH to a healthy level between 7-8 since acids have a pH of less than 7. Water is considered neutral at a pH of 7 and basic above 7. Most ponds tend to be slightly basic. I took the pH of the pond during homeroom today and it was 7.5."

"Thanks for the explanation and for checking the pH today," said Mr. Hull.

The students were anxious to begin their next project using the pond.

"Just remember, students, there is a lot of science to learn from ponds. Enjoy the weekend and I will see all of you next week. Don't forget to bring in samples of water and to follow the protocol I showed you to cap the vial under water so it does not get contaminated from the oxygen in the air or the dissolved oxygen test would not be accurate.

Madison, Savannah, Owen and Jayden were already making plans to meet and to gather samples of water from their neighborhoods.

Discussion Questions

1. What is a wetland and why should it be protected.
2. Why was it important to place indigenous organisms in the pond?
3. Describe or illustrate a small food chain in the pond.
4. Describe or illustrate a small food web in the pond.
5. Describe or illustrate a small food pyramid.
6. Explain why using models such as food chains, webs and pyramids provides a better understanding of an ecosystem.
7. Why was it important to monitor the pH of the pond?
8. What was the purpose of tagging the Monarch Butterfly?
9. Why is milfoil weed and purple loosestrife considered invasive species?
10. Why is it important to monitor the chemistry of water?

Project Pond-Wetland Science Terms

1. *Acid*: A substance with a pH of less than 7.
2. *Amoeba*: A single celled organism that moves by forming pseudopods or false feet.
3. *Bacteria*: Single-celled organisms. Some cause decay.
4. *Base*: A substance with a pH of more than 7.
5. *Ecosystem*: How living and nonliving things interact within an environment.
6. *Chrysalis*: An immature stage of a butterfly when it changes into an adult.
7. *Consumer*: Any organism that eats no longer exists such as the dinosaurs.
8. *Decay:* The process by which microbes break down dead matter.
9. *Euglena*: A single celled organisms that moves with flagella and carries out the life activity of photosynthesis.
10. *Flatworm*: A simple worm
11. *Flow of Energy*: The movement of energy in an ecosystem from one organism to another.
12. *Food Chain*: A model that displays a simple relationship among different organisms in an ecosystem.
13. *Food Relations*: How living things interact to obtain needed nourishment.
14. *Food Pyramid*: A model that displays energy relationships in an ecosystem.
15. *Food Web*: A model that displays more than one way an organism interacts in its environment.
16. *Indigenous*: When an organism is native to a particular region.
17. *Invasive Species*: Organisms foreign to a region and interfere with the life activity of indigenous organisms.
18. *Larva*: An immature stage of an insect such as a caterpillar. This stage often causes the most damage to the environment since the larva is actively eating.
19. *Limnology*: The study of water.
20. *Mating:* The process by which organisms exchange genetic material to make new *Organisms*: Living things that carry out life activities.
21. *Migrate:* To move from one area to another as the seasons change.
22. *Milfoil Weed*: An invasive plant species found in many bodies of water in the northeast of the United States.

23. ***Milkweed***: A plant that is the primary food of monarch butterfly larva.
24. ***Neutral***: A substance with a pH of 7.
25. ***Paramecium***: A single-celled organism that moves with cilia.
26. ***pH***: A measurement of the acidity or basicity of a substance.
27. ***Producer***: An organism that makes its own food. Plants are producers.
28. ***Purple Loosestrife***: An invasive species found in many wetlands on the northeast of the United States.
29. ***Wetland***: A region that stays wet for a period of time.

NGSS ADDRESSED

MS-LS2 Ecosystems: Interactions, Energy, and Dynamics
Interdependent Relationships in Ecosystems
1. Organisms, and populations of organisms, are dependent on their environmental interactions both with other living things and with nonliving factors.
2. In any ecosystem, organisms and populations with similar requirements for food, water, oxygen, or other resources may compete with each other for limited resources, access to which consequently constrains their growth and reproduction.
3. Growth of organisms and population increases are limited by access to resources.
4. Similarly, predatory interactions may reduce the number of organisms or eliminate whole populations of organisms. Mutually beneficial interactions, in contrast, may become so interdependent that each organism requires the other for survival. Although the species involved in these competitive, predatory, and mutually beneficial interactions vary across ecosystems, the patterns of interactions of organisms with their environments, both living and nonliving, are shared.

Cycle of Matter and Energy Transfer in Ecosystems
1. Food webs are models that demonstrate how matter and energy is transferred between producers, consumers, and decomposers as the three groups interact within an ecosystem. Transfers of matter into and out of the physical environment occur at every level. Decomposers recycle nutrients from dead plant or animal matter back to the soil in terrestrial environments or to the water in aquatic environments. The atoms that make up the organisms in an ecosystem are cycled repeatedly between the living and nonliving parts of the ecosystem.
2. Biodiversity describes the variety of species found in Earth's terrestrial and oceanic ecosystems. The completeness or integrity of an ecosystem's biodiversity is often used as a measure of its health.

Biodiversity and Humans
1. Changes in biodiversity can influence humans' resources, such as food, energy, and medicines, as well as ecosystem services that humans rely on—for example, water purification and recycling.

Developing Possible Solutions
1. There are systematic processes for evaluating solutions with respect to how well they meet the criteria and constraints of a problem.

Students who demonstrate understanding can:
1. Analyze and interpret data to provide evidence for the effects of resource availability on organisms and populations of organisms in an ecosystem.
2. Construct an explanation that predicts patterns of interactions among organisms across multiple ecosystems.
3. Develop a model to describe the cycling of matter and flow of energy among living and nonliving parts of an ecosystem.
4. Construct an argument supported by empirical evidence that changes to physical or biological components of an ecosystem affect populations.
5. Evaluate competing design solutions for maintaining biodiversity and ecosystem services.

Joan Wagner has a love affair with science education. She taught science for 34 years, grades 7-12, mostly in grades 7&8. She is the co-author of the *Big 8 Science Review and Test Prep*, published by N&N Publishing and had written 3 workbooks for DK Publishing called "Learn Science," which adapted UK books for the US. She is a past president of the Science Teachers Association of New York State (STANYS) and is presently a member of its conference committee. She chairs the Education Committee for the Dudley Observatory and is the Director of the Greater Capital Region Science and Engineering Fair, an affiliate of the Regeneron International Science and Engineering Fair.

Tales of Science is written to teach science through story-telling and encourage children, ages 12-14, to read science. The book is closely articulated to the *Next Generation of Science Standards*. It includes a number of activities students can do at home or school. *Tales of Science* can be used to support a textbook or even in lieu of a textbook.

Thanks to Dr. Valerie Rapson, former Director of the Dudley Observatory for her astronomy suggestions, Dr. Jennifer Francis for her suggestions of climate change and weather, and my granddaughter, Emilia Wagner for her proofreading.

View of Adirondack Mountains from Cobble Lockout

www.ingramcontent.com/pod-product-compliance
Lightning Source LLC
Chambersburg PA
CBHW080452220526
45465CB00006B/2244